214

SFR 62

William A. Jury
Kurt Roth

Transfer Functions and Solute Movement through Soil

Theory and Applications

1990

Birkhäuser Verlag
Basel · Boston · Berlin

Authors' addresses:

Prof. William A. Jury
Department of Soil and
Environmental Sciences
University of California
Riverside, CA 92521
USA

Kurt Roth
Soil Physics
Institute of Terrestrial Ecology
Eidgenössische Technische Hochschule
Zürich
Switzerland

Library of Congress Cataloging-in-Publication Data

Jury, William A., 1946–
Transfer functions and solute movement through soil : theory and
applications / William A. Jury, Kurt Roth.
p. cm.
Includes bibliographical references and index.
ISBN 0-8176-2509-7 (U.S.)
1. Soils – Solute movement. I. Roth, Kurt, 1955– . II. Title.
S592.3.J87 1990
631.4'3 – dc20

Deutsche Bibliothek Cataloging-in-Publication Data

Jury, William A.:
Transfer functions and solute movement through soil : theory
and applications / William A. Jury ; Kurt Roth. – Basel ;
Boston ; Berlin : Birkhäuser, 1990
ISBN 3-7643-2509-7
NE: Roth, Kurt:

Printed in Germany on acid-free paper
ISBN 3-7643-2509-7
ISBN 0-8176-2509-7

Contents

List of Figures

List of Symbols

		eq.	pg.
Lower Case Latin Symbols			
f_a^f	flux pdf of adsorbing solute	4.12	67
$\widehat{f}$	Laplace transform of f	A.1	151
$\widetilde{f}$	Fourier transform of f	A.49	160
f^f	flux pdf	2.3	24
f_m^f	flux pdf of non-adsorbing solute	4.12	67
f^r	resident pdf	3.37	54
f_{oc}	organic C fraction	4.71	80
ℓ	reference depth for stochastic-convective flux pdf	2.65	40
$\boldsymbol{u}_f$	fluctuation of fluid velocity	7.31	132
$\overline{\boldsymbol{u}}$	mean component of fluid velocity	7.31	132
Upper Case Latin Symbols			
C_a^r	mass of solute adsorbed per mass of dry soil	4.8	66
$\mathbf{C}$	random process of the concentration in a stochastic stream tube model	4.1	64
$\overline{C}$	ensemble average of $\mathbf{C}$	4.1	64
C^f	flux concentration	3.1	45
C_{im}^r	immobile phase resident concentration	3.16	50
C_l^r	mass of dissolved solute per volume of fluid	4.8	66
$\widehat{C}_l^r$	double Laplace transform of C_l^r	D.97	186
C_m^f	flux concentration in mobile phase	3.15	48
C_m^r	mobile phase resident concentration	3.15	48

C_t^r	total resident concentration	3.2	46
I	net applied water	2.5	25
K_d	distribution coefficient	4.9	66
K_{oc}	organic C partition coefficient	4.71	80
J_w	area-averaged steady water flux rate	2.15	27
J_s	solute mass flux	2.34	31
P_a^f	cumulative travel time distribution function of an adsorbing solute	4.11	67
P_m^f	cumulative travel time distribution function of a non-adsorbing (mobile) solute	4.11	67
R	retardation factor	4.10	66
RMF	residual mass fraction	4.69	79
$\boldsymbol{U}$	random fluid velocity field	7.31	132
W	water stored in profile	6.3	109
$\boldsymbol{X}$	trajectory of a particle	7.33	132
$\boldsymbol{Z}$	stochastic process	7.11	127
$\overline{Z}_n$	nth moment of the stochastic process $\boldsymbol{Z}$	7.12	127
Z_N	Nth depth moment of the travel depth pdf	3.39	54

Lower Case Greek Symbols

$\boldsymbol{\lambda}$	wave vector	A.51	160
λ_z	integral correlation length	7.17	128
ρ_b	soil bulk density	4.8	66
ρ	correlation function	7.16	128
σ	covariance function	7.13	127
$\boldsymbol{\sigma}$	covariance tensor	7.32	132
θ	volumetric water content	2.35	31
θ_{im}	volumetric water content of immobile phase	3.15	48
θ_m	volumetric water content of mobile phase	3.15	48
τ_{in}	solute input time	2.1	23
τ_{life}	solute lifetime	2.1	23

Mathematical Symbols and Functions

$\frac{\partial}{\partial t}$	partial differentiation with respect to time		
$\frac{\partial}{\partial z}$	partial differentiation with respect to depth		
$\mathcal{F}$	Fourier transform operator	A.49	160
$\mathcal{F}^{-1}$	inverse Fourier transform operator	A.50	160
$\mathcal{L}$	Laplace transform operator	A.1	151
$\mathcal{L}^{-1}$	inverse Laplace transform operator	A.31	156
E	expectation or ensemble average	2.27	29
$\langle . \rangle$	expectation or ensemble average	7.11	127
Var	variance	2.30	30
Cov	covariance	7.13	127
CV	coefficient of variation	4.7	65
:=	the left side is *defined* in terms of the right side		
$\Rightarrow$	the left side *implies* the right side		
δ	delta function	2.8	26
erf	error function	3.13	48
erfc	complementary error function	3.12	48
H	Heaviside function	2.13	26

Preface

Man's interest in transport processes through soils is probably as old as his active use of nature as a food supply. This interest began to grow rapidly in the last century as the population density—and therefore the inability to abandon degraded land—increased. More recently, the intense interest in chemical contamination of ground water has greatly expanded the range of applications of solute transport beyond the field of agriculture. Now many disciplines—soil science, hydrology, plant science, civil and environmental engineering to name a few—have areas of research that require understanding solute transport in soil.

Advances in the understanding of physical transport in the nineteenth century by Fourier, Ohm, and others, promoted an analogous development in soil physics in the first half of this century. Through the work of Buckingham, Gardner, Richards, and their colleagues, transport equations were formulated for soil based on differential conservation and flux laws, and therefore on a microscopic description of soil. Although these laws worked out very well in small scale laboratory transport experiments with homogeneous soil, they turned out to be of marginal value in the field regime because of the enormous heterogeneity of natural soils. Currently, there are different approaches being taken to overcome the difficulty posed by this heterogeneity.

In contrast to most other approaches, transfer functions formulate the transport process in terms of an integral property of the soil (the impulse response function) which is defined in terms of an experiment (or equivalently, in terms of the response of the system to a mathematical impulse, the Dirac pulse, for theoretical analyses). This experiment yields an integral description of the soil which incorporates all heterogeneities that influence the transport process.

In developing a treatise on solute transport through soil we chose to describe the process in terms of transfer functions for several reasons. First, the transfer function model is ideally suited for representing soil column effluent concentrations, by far the largest source of solute transport information. Second, since any linear solute transport model can be expressed as a transfer function, the common framework of this description facilitates both the understanding of process assumptions in the various models, and the design of experimental tests to evaluate the hypotheses of different models. Finally, in the field regime, where process-based transport models may have large numbers of parameters that are difficult to measure *in situ*, the transfer function model is less data intensive than other approaches, and readily supports either a deterministic or a stochastic framework.

The theory of transfer functions developed in this book attempts to bridge the gap in a practical manner between linear systems modeling where the formulation is exact, and the soil regime, where water and solute transport properties may be transient and nonlinear. This task is approached pragmatically in many instances in the text, by posing and analysing questions such as: Is the integral transfer function approach fundamental in the sense that it can describe transport through a soil under a wide range of conditions? What relationships exist between the integral description and microscopic model assumptions? What type of experiment needs to be performed in a specific application to extract the largest amount of information about the soil, or to test the assumptions of a specific process model?

The first three chapters of the book deal with the formulation and measurement of transfer functions in soil, and with the relationship between flux and resident concentrations in systems described by transfer functions. Chapters 4 and 5 cover the application of the method to the description of transport through soil which has spatially variable transport and reaction properties normal to the flow direction, and to transport through soil which is heterogeneous along the direction of flow. Chapter 6 covers the practical application of the transfer function approach to transport through natural field soils, including its extension to transient water flow conditions. The final Chapter 7 discusses the relation between the transfer function approach and stochastic continuum modeling.

The book contains over 35 worked examples in the text, and an additional 60 solved problems in an appendix. It should be suitable for an advanced undergraduate or graduate level course on solute transport through porous media, or for a reference book for hydrologists, soil scientists, civil or environmental engineers.

Many people have contributed to the development and improvement of this book. Much of the foundational work arose in connection with collaborations between the senior author and Professors Garrison Sposito of UC Berkeley, Robert White of Massey University, and Greg Butters of Colorado State University. Many of the topics covered in the book first arose in a weekly discussion group at UC Riverside consisting of the senior author, and Professors Eileen

Kladivko and Tim Ellsworth, and Drs. Jens Utermann, Jeremy Dyson and Jochim Gruber.

The book was actually written while the senior author was on sabbatical leave at the Eidgenössische Technische Hochschule in Zürich. Appreciation is expressed to Mr. Andreas Papritz, Mr. Thomas Gimmi, Ms. Sabine Koch, and Mr. Martin Schneebeli for helpful comments on the improvement of the text and problems. Professor Rainer Schulin, and Drs. Bernhard Buchter and Mischa Borkovec made many valuable suggestions, and also found a number of mistakes in earlier drafts of the material. A special debt of gratitude is owed to Mr. Markus Flury, who carefully read the final draft and reworked all examples and problems, detecting a number of errors that had eluded earlier attempts at editing.

Above all, the authors are grateful to Professor Hannes Flühler of ETH, without whose encouragement, support, and participation this book could not have been written.

Zürich, June 1990

William A. Jury
Kurt Roth

Chapter 1

Introduction and Philosophy

Many flow systems in nature have characteristic input and output boundaries through which a quantity of interest (e.g. water, heat, a chemical species) enters and leaves the volume where transport and transformation processes occur. The internal dynamics of such a system might be extremely complicated, involving interacting processes with many different characteristic time scales. Modeling the space and time dependence of the quantity of interest within the transport volume may therefore be an insurmountable task, both because of uncertainty about the exact description of the transport or reaction process, or because the properties of the transport volume which are represented in the model process description are not accessible to measurement.

However, if the outflow rate from the transport volume is the only feature of the transport process which must be characterized, a transfer function model can be used as an alternative to describing the system with an internal process model. Transfer functions are used to model complex systems in a simple way by characterizing the output flux as a function of the input flux. The transformation of an arbitrary input signal into an output signal for a linear system is achieved by means of the impulse response function, which defines the response of the system to a narrow pulse input at the inlet end (Himmelblau, 1970).

An example of how one might use a transfer function to model a complex problem is illustrated in Figure 1.1, which shows a chemical mixing vessel agitated by constantly spinning propellers in the tank, mixing two constant (time-independent) volume flows, one containing solute (Q_s) and the other salt-free fresh water (Q_f). The saline concentration of the inlet end varies as a function of time, and the mixing process within the vessel is imperfect (i.e. the concentration varies with position within the vessel). The task of the modeler is to predict the outflow concentration as a function of time for an arbitrary time-dependent concentration $C_s(t)$ entering the saline inflow port, assuming that all solute which enters the inflow end will exit the outflow end (i.e. no precipitation).

Figure 1.1: A chemical mixing vessel with a saline entry port of volume flux Q_s and concentration C_s mixing with a fresh water ($C_f = 0$) port of volume flux Q_f. The outflow volume $Q_o = Q_s + Q_f$ is constant.

In general, this would be an extremely difficult problem to solve. The turbulent motion of the fluid inside the tank cannot be predicted with precision, and therefore the convection and dispersive mixing of the saline fluid as it moves through the vessel toward the outflow end cannot be modeled in terms of the observable properties of the vessel, such as its geometry, and the measurable inflow volumes and concentrations.

However, the outflow concentration is the only characteristic of this system which we wish to describe with our model, and all of the processes which can influence the transport and mixing of a solute molecule during its transit through the vessel are the same from one moment to the next. Consequently, if a large number of solute molecules are added at one instant ($t = 0$) to the saline port, and thereafter both ports deliver saline-free water to the mixing vessel, the outflow concentration for $t > 0$ contains all the information required to calculate a travel time probability distribution or probability density function (pdf) for the solute molecules. The travel time pdf characterizes the distribution of possible travel times that a solute molecule might experience in moving from the inlet end to the outlet end. The connection between the outflow concentration and the pdf may be drawn as follows. Suppose that N solute molecules are added suddenly to the saline inflow stream over the time period $0 < t < \Delta t$, where Δt is very small. At the outflow end, solution is collected over time intervals $\Delta t_j = t_j - t_{j-1}$ and the number of solute molecules recovered in each interval is estimated from the measured concentrations. The number of molecules n_j recovered during the jth time interval $[t_{j-1} < t < t_j]$ divided by the total number N added is thus approximately equal to the probability that a single solute molecule has a travel time in this interval. As long as the number of molecules added are large enough to sample all of the possible flow paths, and the interval of time over which they are added is much shorter than the shortest travel time through the system, then we may write

$$\frac{n_j}{N} \approx P(t_j) - P(t_{j-1}) \approx f(t_{j-\frac{1}{2}}) \Delta t_j \tag{1.1}$$

where $P(t)$ is the cumulative travel time distribution function (cdf), defined as the probability that a solute molecule added at $t = 0$ will have a travel time that is less than or equal to t, and $f(t)$ is the corresponding probability density function.

Since the system is operating under steady state (time-independent) water flow conditions, n_j is proportional to the average solute outflow concentration during the interval of measurement. Thus,

$$C_o(t_{j-\frac{1}{2}}) = \frac{\mu n_j}{Q_o \Delta t}, \tag{1.2}$$

where μ is the solute mass per molecule and $Q_o = Q_s + Q_f$ is the outflow volume flow rate. Thus, in the limit $\Delta t \to 0$, (1.1) and (1.2) imply that[1]

$$f(t) = \frac{C_o(t)}{\int_0^\infty C_o(t)\, dt} = \frac{C_o(t)}{\mu N / Q_o}. \tag{1.3}$$

The graphical counterparts of (1.1) and (1.3) are shown in Figure 1.2.

Equation (1.3) states that the travel time pdf of the solute molecules entering the saline port and leaving the outlet port is equal to the normalized (unit area) outflow concentration-time curve when a narrow pulse of solute is added to the

[1] Note that $f(t)$ has dimensions of time^{-1}, because $f(t)\,\Delta t$ is dimensionless.

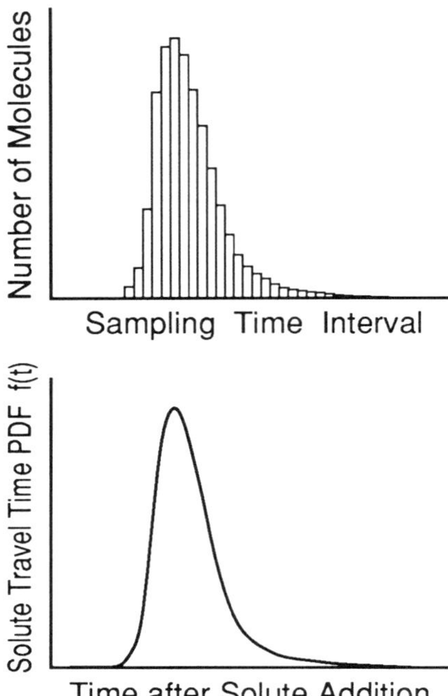

Figure 1.2: *Relationship between the solute travel time pdf and the measurement of solute molecules in the outflow volume.*

inlet end. This function $f(t)$ is sometimes called the impulse response function of the system (Himmelblau, 1970).

When an arbitrary inlet concentration $C_s(t)$ is present in the saline volume flow over time, we assume that a given molecule entering at a particular time will move through the system randomly while being governed by the same probability laws as a different molecule entering the system at any other time. This assumption is based on the fact that none of the forces driving the solute through the system are time-dependent. Therefore, a solute molecule which enters the vessel at time $t - t'$ has a probability $f(t')\,dt'$ of having a travel time between t' and $t' + dt'$, or equivalently, has a probability $f(t')\,dt'$ that it will leave the vessel at time t (Jury et al., 1986). The outflow concentration at time t is thus equal to

$$C_o(t) = \int_0^t C_s(t - t')\, f(t')\, dt' = \int_0^t C_s(t')\, f(t - t')\, dt' \ , \tag{1.4}$$

where the last form of the equation results from a change of variable. Equation (1.4) is simply an application of the principle of linear superposition. It states that the outflow concentration at time t is made up of contributions from solute molecules that entered the inflow end at various earlier times $t - t'$, multiplied the probability $f(t')\,dt'$ of having a travel time t'. This equation, called a convolution integral (Arfken, 1985), is the transfer function model of the outflow concentration.

Some advantages of the transfer function approach are immediately obvious. The internal dynamics of the system are characterized implicitly by one single measurable property, the impulse response function $f(t)$. With this function, the observed input signal $C_s(t)$ is transformed into the output signal $C_o(t)$ with the convolution integral or transfer function model (1.4). No process model of solute transport within the vessel is needed.[2]

Clearly, there are several assumptions required for this approach to work. Some of them (i.e. steady state, nonreactive solute) were introduced specifically for the mixing vessel example. However, even when the system is constrained by the set of assumptions used in this example, there are other questions which arise about the validity of the model. For instance, under what conditions does the principle of superposition embodied in the convolution integral (1.4) hold for solute transport? Can the flow volumes Q_s and Q_f be time dependent?

The answers to these and similar questions will be forthcoming only after a more general development of the transfer function model of solute transport is undertaken, and this task will form the major part of the book.

Problems

Problem 1.1 Develop a process model of the transfer function travel time pdf for the system shown in Figure 1.1 for the special case of infinitely fast and perfect mixing inside the vessel. Assume that the volume of water in the tank is always equal to V.

Problem 1.2 Develop a general transfer function solution for the perfectly mixed tank in Problem 1.1 when the saline concentration is $C_s(t)$. Indicate what changes must be made in this solution when the saline flow rate Q_s is time dependent (but Q_o is constant).

Problem 1.3 Derive the last form of the convolution integral (1.4) from the first one. Discuss the physical meaning of this form of the integral.

---

[2]In a sense, $f(t)$ is the process model, since it describes the travel time distribution.

Problem 1.4 Suppose that some aspect of the solute transport process through the mixing vessel was a nonlinear function of the local solute concentration. Could the transfer function convolution integral be used in this case?

Chapter 2

The Solute Transfer Function for Transport through Soil

2.1 The Solute Transport Volume

The extension of the transfer function approach to solute transport through soil is straightforward. We begin by picturing a volume V of soil (Figure 2.1) bounded externally by a surface S, the top of which intersects the soil surface. The lateral extent of V is either constrained by walls (as in a soil column) or is chosen to be large enough that no solute enters or leaves through it during the period of monitoring.[1] Inside of V is a smaller volume called the solute transport volume V_{ST}, which includes all of the fluid volume which is effective in transporting solute during the period of observation.

This interior volume V_{ST} is bounded by a complex surface S_{ST} which forms the boundary between the fluid that is effective in transporting solute, and the rest of the soil volume, which might include stagnant fluid phases that are not assisting in the transport process (Jury et al., 1986).

In general, both V_{ST} and S_{ST} can be time-dependent. Further, the definition of V_{ST} is somewhat arbitrary, and will require an operational interpretation of the phrase "effective in transporting solute" in terms of observable characteristics of the solute transport process. This attribute of the soil solution has been represented in convective-dispersive models of solute transport through porous

[1] This requirement can be relaxed somewhat to allow no net inflow or outflow along the lateral boundaries. This extension would allow an interior part of a large volume like an agricultural field which has a spatially uniform upper boundary condition to be treated as a transport volume, even though fluctuating components of lateral flow could occur at various points along the periphery.

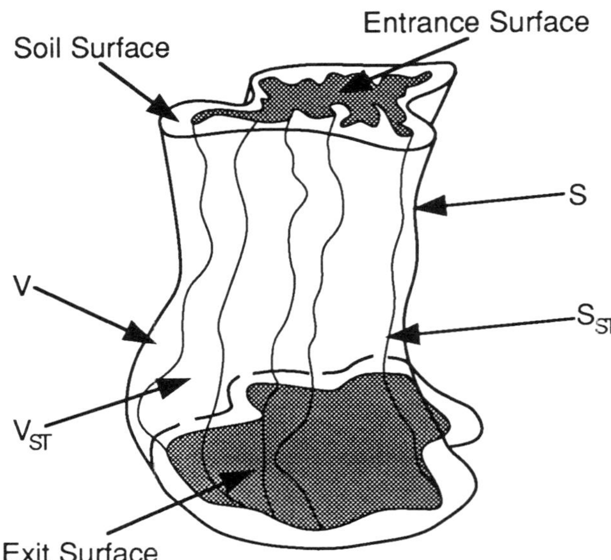

Figure 2.1: *Schematic illustration of a field soil unit of volume V bounded by a surface S and enclosing a transport volume V_{ST} and its bounding surface S_{ST} (adapted from Jury et al., 1986).*

media as a "mobile water volume" (Coats and Smith, 1956; Van Genuchten and Wierenga, 1976) or as a "mobile water zone" (Addiscott, 1977), where it specifically refers to the fraction of the wetted soil volume through which water, containing dissolved solute, is moving. In this discussion, however, no reference is made to specific mechanisms of transport, except for the idea that not all of the fluid in the soil matrix need play an active role in transport of solute.

Conceptually, solute can enter the transport volume V_{ST} in four distinct ways: (i) it may arrive through the entrance surface by controlled or accidental application to the soil surface; (ii) it may move from the soil solid phases into the transport volume by dissolution or desorption processes; (iii) it may move from the gaseous phase into solution; or (iv) it may appear in the transport volume from chemical, physical or biological processes which occur within it. Similarly, there are four distinct means by which solute can leave the transport volume: (i) exit through the lower boundary, (ii) movement to the solid phase by precipitation or adsorption; (iii) movement to the vapor phase by volatilization, or (iv) disappearance by transformation to a different form.

Regardless of the mechanisms which are involved, it is possible to define two important time variables which characterize a solute molecule in the transport volume V_{ST} (Jury et al., 1986): (i) the *solute input time* τ_{in} (the time when it first appears in the volume by any of the means discussed above), and the *solute lifetime* τ_{life} (the period of time it stays in the volume). The input time and lifetime are common elements of chemical reaction engineering models, some of which are transfer functions (Himmelblau and Bischoff, 1968; Naumann and Buffham, 1983).

The generalization of the transfer function equation derived in Chapter 1 to describe solute transport through V_{ST} under the general conditions discussed above is (Jury et al., 1986)

$$Q_{out}(t) = \int_0^t g(t - \tau_{in} \mid \tau_{in})\, Q_{in}(\tau_{in})\, d\tau_{in} \ , \tag{2.1}$$

where $Q_{in}(t)$ is the rate at which solute mass is entering the transport volume, $Q_{out}(t)$ is the rate at which it is leaving, and $g(\tau_{life} \mid \tau_{in})$ is the solute lifetime pdf, describing the distribution of solute lifetimes τ_{life}, conditional on the time τ_{in} at which solute enters V_{ST} (Jury et al., 1986; Sposito and Jury, 1988). The lifetime pdf is a model representation of the net effect of all the processes that can influence the time a solute molecule spends in the soil. It is conditional on the input time τ_{in} to allow for the possibility that the physical, chemical, or biological processes that can affect the lifetime τ_{life} of a solute molecule may be time-dependent.

Although (2.1) is general enough to represent any solute transport process that is linear,[2] it is not practical for several reasons. First, internal modes of entry or exit such as dissolution/precipitation reactions or biochemical transformation are normally not directly measurable, so that the input and output flow rates cannot be completely monitored when solute enters or leaves V_{ST} other than through the external surfaces. Second, if $g(\tau_{life} \mid \tau_{in})$ is conditional on the input time, it must first be measured for all input times before (2.1) can be used to calculate $Q_{out}(t)$.

For these reasons, (2.1) should be regarded as a formal relation, which connects the solute transport process with the laws of probability governing the transport and transformation processes a solute molecule may encounter in the solute transport volume. However, there are important special cases of this equation which are practical, and which can serve in lieu of a comprehensive theory of solute transport through the soil. Moreover, all mechanistic linear models of solute transport are consistent with the formulation in (2.1), and therefore representations of $g(\tau_{life} \mid \tau_{in})$ may be generated from specific process hypotheses. (See Rinaldo et al., (1989) for an example of such a model.)

[2]The requirement that must be satisfied for (2.1) to be valid is that if $Q_{in}(t) = \alpha Q_{1_{in}}(t) + \beta Q_{2_{in}}(t)$ then $Q_{out}(t) = \alpha Q_{1_{out}}(t) + \beta Q_{2_{out}}(t)$, where $Q_{1_{out}}$ and $Q_{2_{out}}$ are the outflows corresponding to $Q_{1_{in}}$ and $Q_{2_{in}}$, respectively.

2.2 The Travel Time PDF

If (2.1) is applied to the special case of a conservative (nonvolatile, nonreactive and nonadsorbing) solute, then the only modes of entry into and exit from the transport volume are through the upper and lower exterior surfaces. In this case, the input and output mass flow rates in (2.1) are now the solute mass flow rates (flux · area) at the inlet and outflow ends of V, respectively, both of which can be monitored. Also, the solute lifetime τ_{life} now becomes a travel time describing the amount of time required for a solute molecule to leave the transport volume through the outflow end after entering it through the inflow end. If the surface areas A of the inflow and outflow boundaries are the same size, then we may introduce the solute concentration into the transfer function equation by the substitutions

$$\begin{aligned} Q_{in}(t) &:= Q(0,t) = Q_w(0,t)\,C^f(0,t) = AJ_w(0,t)C^f(0,t) \\ Q_{out}(t) &:= Q(\ell,t) = Q_w(\ell,t)\,C^f(\ell,t) = AJ_w(\ell,t)C^f(\ell,t)\ , \end{aligned} \tag{2.2}$$

where Q_w is the soil water volume flow rate, $J_w = Q_w/A$ is the water flux, and C^f is the area-averaged solute flux concentration (Kreft and Zuber, 1978; Parker and Van Genuchten, 1984). The soil volume represented in Fig. 2.1 is now assumed to be vertical with entrance and exit surfaces at $z = 0$ and $z = \ell$, respectively. The solute flux concentration C^f is the ratio of the solute mass flow rate to the water flow rate. Its relation to the spatial solute mass density or solute resident concentration C^r will be developed in detail in Chapter 3.

Substitution of (2.2) into (2.1) produces the following version of the transfer function equation

$$C^f(\ell,t) = \int_0^t \frac{C^f(0,\tau_{in})\,J_w(0,\tau_{in})}{J_w(\ell,t)}\,f^f(\ell,t-\tau_{in}\mid\tau_{in})\,d\tau_{in} \tag{2.3}$$

where $f^f(\ell,\tau \mid \tau_{in})$ is the travel time pdf,[3] which is identical to the solute lifetime pdf $g(\ell,\tau \mid \tau_{in})$, for a conservative tracer. The travel time pdf $f^f(\ell,\tau \mid \tau_{in})$ is still conditional on the solute input time τ_{in} because the water flow processes occurring within the transport volume may be time-dependent. Since it will be shown later that the travel time pdf is a flux concentration, it will be given a superscript f from now on.

Systems that need to be described by (2.3) are still difficult to characterize, because the dependence of $f^f(\ell,\tau \mid \tau_{in})$ on the input time τ_{in} must be measured

[3] In this notation, τ is a random variable and ℓ is a parameter.

or explicitly modeled. This requirement will be addressed in several ways in the book, and in considerable detail in Chapter 6. For the special case of steady water flow, J_w is a constant, and the travel time pdf is no longer conditional on the input time ($f^f(\ell, t-\tau_{in} \mid \tau_{in}) = f^f(\ell, t-\tau_{in} \mid 0) = f^f(\ell, t-\tau_{in})$). This simplification reduces (2.3) to

$$C^f(\ell,t) = \int_0^t C^f(0,\tau_{in})\, f^f(\ell, t-\tau_{in})\, d\tau_{in} \tag{2.4}$$

Examples of solute transport volumes where (2.3) or (2.4) might be used are soil columns, field areas receiving water and chemical input fluxes over the entire inlet surface, such as in agriculture, and tile drained fields. In the last case, however, the outflow surface is merely the drain tube, while the inlet surface is the entire soil area. Therefore the mass flux form of (2.3) or (2.4) should be used.

During steady flow, the cumulative net applied water- or drainage flux $I = J_w t$ is proportional to time. Therefore, (2.4) may be rewritten as

$$C^f(\ell,I) = \int_0^I C^f(0,I') f^f(\ell, I-I')\, dI' \ . \tag{2.5}$$

Equation (2.5) is called the net applied water form of the transfer function equation (Jury, 1982). It will be shown in Chapter 6 that this form of the model, with suitable adjustments, may be used to describe solute transport under transient water flow for certain soils that do not change their solute transport characteristics appreciably when the water flux changes.

Excursus THE DELTA FUNCTION $\delta(t)$ AND THE HEAVISIDE FUNCTION $H(t)$

Many of the experiments conducted to study solute transport through soil involve adding pulses or step changes of solute to the inlet end. These two operations can be represented mathematically with generalized functions called the *Dirac delta function* $\delta(t)$ and the *Heaviside unit function* $H(t)$.[4]

To define the delta function, first consider the Gauss function $N(t;\sigma)$ with mean zero and variance σ^2, which is given by

$$N(t;\sigma) := \frac{1}{\sqrt{2\pi}\sigma} \exp\Big(-\frac{t^2}{2\sigma^2}\Big) \tag{2.6}$$

As long as $N(t;\sigma)$ is well defined, the integral

$$\int_{-\infty}^{+\infty} N(t;\sigma)\, dt = 1 \tag{2.7}$$

[4]Generalized functions naturally occur as limiting cases of normal functions. An example is the Gauss function in the limit of vanishing variance.

is independent of the variance σ^2. The delta function $\delta(t)$ may be defined as the limit of the Gauss function as the variance approaches zero,[5]

$$\delta(t) := \lim_{\sigma \to 0} N(t;\sigma) \ . \tag{2.8}$$

This generalized function has the remarkable property that $\int_{-\infty}^{+\infty} \delta(t)\, dt = 1$—by extension of (2.7)—while $\delta(t) = 0$ for all $t \neq 0$—from (2.6) and (2.8). It follows from these properties that the integral over the product of $\delta(t-a)$ and a continuous function $f(t)$ yields the value of f at the point a where the argument of δ is zero,

$$\int_{-\infty}^{+\infty} f(t)\, \delta(t-a)\, dt = f(a) \ , \tag{2.9}$$

which is sometimes called a filtering property. Equation (2.9) may be generalized to include functions in the argument of the delta function. This generalization may be expressed as

$$\int_{-\infty}^{+\infty} f(t)\, \delta(g(t)-a)\, dt = \int_{-\infty}^{+\infty} \frac{f(g^{-1}(\tau))\, \delta(\tau - a)}{\frac{dg}{dt}(g^{-1}(\tau))}\, d\tau = \frac{f(g^{-1}(a))}{\frac{dg}{dt}(g^{-1}(a))} \ , \tag{2.10}$$

where $g(t)$ is any invertible function. (Equations (2.9)–(2.10) may be verified by using the definition (2.8) of the delta function.)

We will find the delta function very useful for representing the shape of the narrow input pulse that is added to produce the impulse response function at the outlet end of the transport volume.

The Heaviside function $H(t)$ is defined as the limit of the integral

$$P(t;\sigma) := \int_{-\infty}^{t} N(\tau;\sigma)\, d\tau = \frac{1}{\sqrt{2\pi}\sigma} \int_{-\infty}^{t} \exp\Big(-\frac{\tau^2}{2\sigma^2}\Big)\, d\tau \tag{2.11}$$

for vanishing σ,

$$H(t) := \lim_{\sigma \to 0} P(t;\sigma) \ . \tag{2.12}$$

It is easy to verify from (2.11)–(2.12) that

$$H(t) := \begin{cases} 0 & \text{if } t < 0 \\ 1 & \text{if } t > 0 \end{cases} \tag{2.13}$$

by noting that for small σ, the function $N(t;\sigma)$ has large values only within a very small interval around $t = 0$. The delta function may be defined as the derivative of the Heaviside function,

$$\delta(t) = \frac{dH(t)}{dt} \ , \tag{2.14}$$

by comparing the definitions (2.8) and (2.12). The Heaviside function is useful for representing the constant boundary condition which begins at $t = 0$, as well as any other condition which changes abruptly from one value to another.

---

[5]The delta function $\delta(t)$ was originally introduced by P.A.M. Dirac (1947) in the early days of quantum mechanics. The mathematical theory that deals with such objects was developed by L. Schwartz (1950). (See Gel'fand and Shilov (1964) for a modern treatment of the subject.)

2.3 Measurement of the Travel Time PDF $f^f(\ell, t)$

Equation (2.4) suggests a general method for measuring the travel time pdf $f^f(\ell, t)$ of a solute transport volume bounded above at $z = 0$ by an entrance surface through which water and chemicals are added, and below at depth $z = \ell$ by an exit surface where the flux concentration $C^f(\ell, t)$ is monitored.[6] If the water flow regime is at steady state, the solute input flux from a narrow solute pulse of total mass M applied uniformly to the surface of the transport volume over a short time period may be represented as

$$J_s(0, t) = J_w C^f(0, t) = MN(t; \Delta t) , \tag{2.15}$$

where J_w is the area-averaged steady water flux rate through the entrance surface, J_s is the area-averaged solute mass flux, and $N(t; \Delta t)$ is the Gauss function (2.6) with variance Δt^2. In the limit as Δt becomes very small, the Gauss function may be approximated by the delta function $\delta(t)$, and (2.15) reduces to

$$\lim_{\Delta t \to 0} C^f(0, t) = \frac{M}{J_w} \delta(t) . \tag{2.16}$$

If (2.16) is substituted into (2.4), the transfer function integral reduces to (using (2.9))

$$C^f(\ell, t) = \frac{M}{J_w} f^f(\ell, t) . \tag{2.17}$$

Thus, the travel time pdf $f^f(\ell, t)$ is proportional to the outflow flux concentration in response to a narrow pulse input of solute. For this reason, $f^f(\ell, t)$ is sometimes called an *impulse response function.* Thus, by (2.17), we may construct the travel time pdf from

$$f^f(\ell, t) = \frac{C^f(\ell, t)}{\int_0^\infty C^f(\ell, t')\, dt'} . \tag{2.18}$$

Several points should be made about (2.18). First, this definition of the travel time pdf is valid only for a transport volume in which steady water flow is

[6] Since the flux concentration is the ratio of the solute mass flow rate to the water flow rate, the monitoring devices should ideally accept solution passively out of the moving fluid at the exit surface, without disturbing the flow regime in any way. The degree to which this requirement is met by conventional solution monitoring devices such as vacuum solution samplers or ground water extraction wells is not clear.

occurring. Under transient water flow, the travel time pdf is conditional on the input time and (2.3) must be used to interpret the outflow from a narrow pulse inflow experiment. Second, it is clear from (2.18) that $f^f(\ell, t)$ is a normalized (unit area) flux concentration, which is associated with a fixed position ℓ in space where the solute outflow is monitored. Finally, the normalization requires integrating over all finite t. As an alternative, the denominator of (2.18) could be replaced by M/J_w for a conservative solute.

2.4 Measurement of the Travel Time CDF $P^f(\ell, t)$

If the inflow solution concentration at the entrance surface of the transport volume under steady water flow is suddenly changed from 0 to C_0 at $t = 0$, we may express the input flux as

$$J_s(0,t) = J_w\, C^f(0,t) = C_0\, J_w\, H(t) \ . \tag{2.19}$$

Solving for the input flux concentration and inserting it into (2.4), we obtain

$$C^f(\ell,t) = C_0 \int_0^t f^f(\ell,t')\, dt' = C_0\, P^f(\ell,t) \ , \tag{2.20}$$

where $P^f(\ell, t)$ is called the cumulative travel time distribution function or cdf, which is equal to the probability that the travel time will be less than or equal to t. Equation (2.20) suggests another way of estimating the travel time pdf $f^f(\ell, t)$ using the step change input of solute,[7] namely

$$f^f(\ell,t) = \frac{dP_\ell(t)}{dt} = \frac{1}{C_0}\frac{dC^f(\ell,t)}{dt} \ . \tag{2.21}$$

This equation also establishes the pdf as the derivative of the cdf.

Excursus PROPERTIES OF PROBABILITY DISTRIBUTIONS

The cumulative distribution function or cdf $P(t)$ of a random variable $t \geq 0$ is defined in terms of probability concepts as

$$P(t_0) := \text{Prob}\{t \leq t_0\} \ , \tag{2.22}$$

[7] In general, however, this would be less accurate than the pulse input method since it requires taking the derivative of a function characterized only at discrete times.

where t_0 is any value of t between 0 and ∞. Thus, it follows that

$$\begin{aligned} P(0) &= 0 , \\ P(\infty) &= 1 . \end{aligned} \tag{2.23}$$

The probability density function or pdf $f(t)$ is defined as the derivative of the cdf,

$$f(t) := \lim_{\Delta t \to 0} \frac{P(t + \Delta t) - P(t)}{\Delta t} = \frac{dP(t)}{dt} , \tag{2.24}$$

or, in terms of probability concepts

$$f\left(t_0 + \frac{\Delta t}{2}\right) \Delta t \approx \text{Prob}\{t_0 \leq t < t_0 + \Delta t\} . \tag{2.25}$$

Using (2.23) we obtain the normalization condition

$$\int_0^\infty f(t)\, dt = 1 . \tag{2.26}$$

Some properties of the pdf may be defined through its moments.

Mean or Expected Value of t

$$\text{E}(t) := \int_0^\infty t\, f(t)\, dt , \tag{2.27}$$

where E(.) is the expectation operator, which averages the value of its argument over the probability distribution. It is sometimes called an *ensemble average*, because it averages over the ensemble of all possible states of its argument. Equation (2.27) defines the mean value of the random variable t. For example, if $f(t)$ is the travel time pdf $f^f(\ell, t)$, then $\text{E}_\ell(t)$ is the mean travel time between 0 and ℓ.

The mean travel time is one of the fundamental attributes of the travel time pdf, and can be used to calculate the mean solute velocity.

Nth Moment of t

$$\text{E}(t^N) := \int_0^\infty t^N f(t)\, dt , \tag{2.28}$$

If the pdf $f(t)$ is expressed by a functional form that has a known analytic expression $\widehat{f}(s)$ for its Laplace transform (Appendix A), then the Nth moment (2.28) may be calculated by the operation (see Problem 2.9)

$$\text{E}(t^N) = (-1)^N \left. \frac{d^N \widehat{f}(s)}{ds^N} \right|_{s=0} , \tag{2.29}$$

where s is the variable conjugate to t in the Laplace transform operation.

Variance of t

$$\mathrm{Var}(t) := \int_0^\infty (t - \mathrm{E}(t))^2 f(t)\, dt = \mathrm{E}(t^2) - \mathrm{E}^2(t) \tag{2.30}$$

The variance of the random variable t describes the mean square deviation of the values of t about the average value.

The variance of the travel time pdf is one of the important characteristics of solute transport, and will be shown to be related to the solute dispersion coefficient in the classic convective-dispersive solute transport model.

A final expectation operation which will be useful in later parts of the book is the average value of a function $G(t)$ of t.

Expectation of a Function G(t)

$$\mathrm{E}(G(t)) = \int_0^\infty G(t)\, f(t)\, dt \tag{2.31}$$

The properties described above are applicable to any random variable. We will make use of them repeatedly throughout the book.

2.5 Transfer Function Representation of Process Models

To this point, the transfer function approach has been presented as a method of characterizing solute outflow from a soil volume without developing a process model for the transport and transformation processes occurring within it. However, since the assumptions which were used in the development of the travel time form (2.4) of the transfer function may be represented explicitly with a specific process model, travel time representations of process models are also possible. The travel time pdf $f^f(\ell, t)$ has been defined in (2.18) as the normalized outflow concentration corresponding to a delta function input pulse of solute flux. Therefore, specific process representations of $f^f(\ell, t)$ can be derived by using this upper boundary condition. The approach will be explored in several examples below.

Example 2.1 The Piston Flow Model

A soil column of water content θ and length ℓ has water flowing through it in steady state at a flux rate J_w. Chemical dissolved in solution is assumed to move through the column according to the piston flow model; thus, each molecule moves through the column at the same velocity $V = J_w/\theta$ without spreading by diffusion or dispersion. The travel time pdf of this simple system is therefore just a shifted delta function

$$f^f(\ell,t) = \delta(t - \ell\theta/J_w) . \tag{2.32}$$

Equation (2.32) states that all solute molecules that enter the soil column at $t = 0$ have the same travel time, and therefore will exit the outflow end located at depth ℓ at exactly $t = \ell\theta/J_w$. If (2.32) is inserted into the transfer function (2.4), the outflow flux concentration at ℓ for an arbitrary input concentration $C^f(0,t) = C_{in}(t)$ is, using (2.9)

$$C^f(\ell,t) = C_{in}(t - \ell\theta/J_w) . \tag{2.33}$$

Thus, the piston flow model merely produces a time-delayed reproduction of the input signal at any depth ℓ in the soil.

Example 2.2 THE CONVECTION-DISPERSION EQUATION (CDE)

The convection-dispersion model is the most commonly used process representation for representing solute movement through soil (Nielsen and Biggar, 1962; Biggar and Nielsen, 1967). This model assumes that the solute mass flux J_s (mass/area/time) can be written as the sum of two terms: a mass flow term describing the passive convection of dissolved solute within moving soil solution, and a second term describing the random mixing of solute by diffusion and dispersion within the moving fluid.

These two solute flux contributions may be written in terms of the resident fluid concentration C_l^r (mass of dissolved solute/fluid volume) as

$$J_s = -\theta D \frac{\partial C_l^r}{\partial z} + J_w C_l^r , \tag{2.34}$$

where D is the effective diffusion-dispersion coefficient (Biggar and Nielsen, 1967).

The differential equation describing this model is generated by combining (2.34) with the solute conservation equation

$$\frac{\partial \theta C_l^r}{\partial t} + \frac{\partial J_s}{\partial z} = 0 . \tag{2.35}$$

Equation (2.35) is a differential expression of mass conservation for a mobile, nonvolatile solute which does not undergo reactions in the soil. For the case of steady state water flow through a medium with a uniform water content θ, (2.34) and (2.35) may be combined to produce the Convection-Dispersion Equation or CDE

$$\frac{\partial C_l^r}{\partial t} = D\frac{\partial^2 C_l^r}{\partial z^2} - V\frac{\partial C_l^r}{\partial z} , \tag{2.36}$$

where $V = J_w/\theta$ is the so-called pore water velocity or mobile solute velocity. As shown in Parker and Van Genuchten (1984), and also in Chapter 3, (2.36) can also be expressed in terms of the flux concentration

$$\frac{\partial C^f}{\partial t} = D\frac{\partial^2 C^f}{\partial z^2} - V\frac{\partial C^f}{\partial z} . \tag{2.37}$$

This version of the equation is best suited for developing the transfer function form of the model.[8]

To derive the travel time pdf for the CDE model at a depth z, (2.37) must be solved for the boundary and initial conditions which correspond to the measurement of $f^f(z,t)$, i.e.

$$C^f(z,0) = 0 \ , \tag{2.38}$$

$$C^f(0,t) = \delta(t) \ , \tag{2.39}$$

$$C^f(\infty,t) = 0 \ . \tag{2.40}$$

The conditions in (2.38)–(2.40) ensure that no solute is present in the soil initially, and that the integral of the concentration-time curve at any depth will be unity. Therefore, the flux concentration at any depth z will be equal to the travel time pdf for the transport volume between 0 and z.

Equations (2.37) and (2.38)–(2.40) can be solved easily by Laplace transforms, as shown below.

The Laplace transform operation (see Appendix A) applied to (2.37) transforms each term as follows

$$\begin{aligned}\int_0^\infty \frac{\partial C^f}{\partial t} \exp(-st)\,dt \\ &= C^f(z,t)\exp(-st)\Big|_{t=0}^{t=\infty} + s\int_0^\infty C^f(z,t)\exp(-st)\,dt \\ &= s\widehat{C}^f(z;s) - C^f(z,0)\end{aligned}$$

$$\begin{aligned}\int_0^\infty D\frac{\partial^2 C^f}{\partial z^2} \exp(-st)\,dt \\ &= D\frac{\partial^2}{\partial z^2}\int_0^\infty C^f(z,t)\exp(-st)\,dt = D\frac{d^2\widehat{C}^f(z;s)}{dz^2}\end{aligned}$$

$$\begin{aligned}\int_0^\infty V\frac{\partial C^f}{\partial z} \exp(-st)\,dt \\ &= V\frac{\partial}{\partial z}\int_0^\infty C^f(z,t)\exp(-st)\,dt = V\frac{d\widehat{C}^f(z;s)}{dz} \ ,\end{aligned}$$

where

$$\widehat{C}^f(z;s) := \int_0^\infty C^f(z,t)\exp(-st)\,dt \tag{2.41}$$

is the Laplace transform of the flux concentration $C^f(z,t)$; it depends on z and the parameter s. Note that the initial condition (2.38) has become part of the Laplace transform of the time

[8] After a relationship is developed in Chapter 3 between the flux- and resident concentration, it will be possible to use either form of the differential equation to derive the travel time pdf of the model.

derivative of the solute concentration. Therefore, the transformed ordinary differential equation is (using (2.38))

$$D\frac{d^2\widehat{C}^f}{dz^2} - V\frac{d\widehat{C}^f}{dz} - s\widehat{C}^f = 0 \; . \tag{2.42}$$

The boundary conditions (2.39) and (2.40) must also be transformed. Using the fact that the integral of the delta function is unity, the transformed upper boundary condition (2.39) becomes

$$\widehat{C}^f(0) = 1 \tag{2.43}$$

whereas (2.40) yields simply

$$\widehat{C}^f(\infty) = 0 \; . \tag{2.44}$$

Equation (2.42) is an ordinary differential equation with constant coefficients. Solutions of this equation are of the form (Kaplan, 1984)

$$\widehat{C}^f(z) = \exp(mz) \; , \tag{2.45}$$

where m is a constant. When (2.45) is inserted into (2.42), we obtain

$$(D\,m^2 - V\,m - s)\exp(mz) = 0 \; , \tag{2.46}$$

which must be true for all z. Therefore, the quantity in parentheses must vanish, which will happen if

$$m = \frac{V}{2D} \pm \frac{V}{2D}\sqrt{1 + \frac{4sD}{V^2}} = \frac{V}{2D}(1 \pm \xi) \; , \tag{2.47}$$

where

$$\xi := \sqrt{1 + \frac{4sD}{V^2}} \; . \tag{2.48}$$

Thus, the general solution of (2.42) may be written as

$$\widehat{C}^f(z) = A\,\exp\Big(\frac{Vz}{2D}(1-\xi)\Big) + B\,\exp\Big(\frac{Vz}{2D}(1+\xi)\Big) \; , \tag{2.49}$$

where A and B do not depend on z.

Since $\xi > 1$, the only way for the lower boundary condition (2.44) to be satisfied is if $B = 0$. Also, with $B = 0$, (2.49) satisfies the upper boundary condition (2.43) if $A = 1$. Therefore, the solution to (2.42)–(2.44) is

$$\widehat{C}^f(z;s) = \widehat{f}^f(z;s) = \exp\Big(\frac{Vz}{2D}(1-\xi)\Big) \; , \tag{2.50}$$

where $\widehat{f}^f(z;s)$ is the Laplace transform of the travel time pdf $f^f(z;t)$ of the CDE model. The inverse Laplace transform of (2.50) is given in Jury and Sposito (1985). It is also derived in Problem 2.1 using the Table of Laplace Transforms in Appendix C. The solution is

$$C^f(z,t) = f^f(z;t) = \frac{z}{2\sqrt{\pi D t^3}}\exp\Big(-\frac{(z-Vt)^2}{4Dt}\Big) \; . \tag{2.51}$$

By expressing the CDE model solution as a transfer function, we now can write the general solution for the flux concentration at depth z for a time-dependent inflow solute concentration

$C_{in}(t)$ and steady state water flow by inserting the solute travel time pdf (2.51) into the transfer function (2.4)

$$C^f(z,t) = \int_0^t C_{in}(t-t') \frac{z}{2\sqrt{\pi D t'^3}} \exp\Big(-\frac{(z-Vt')^2}{4Dt'}\Big)\, dt' \ . \tag{2.52}$$

Equation (2.52) is now equivalent in every way to the original differential equation (2.37) and the boundary and initial conditions

$$C^f(z,0) = 0 \quad , \quad C^f(0,t) = C_{in}(t) \quad , \quad C^f(\infty,t) = 0 \ . \tag{2.53}$$

In fact, the travel time pdf (2.51) can completely replace the differential equation (2.37) describing the CDE, because it has the same information contained in it.

Once the travel time pdf of a particular process model has been derived by the above procedure, the travel time moments (2.27)–(2.30) can be calculated. This procedure is illustrated in the next two examples for the piston flow model and the CDE.

Example 2.3 TRAVEL TIME MOMENTS OF THE PISTON FLOW MODEL

If we apply the filtering property of the delta function (2.9) to the Nth moment (2.28) of the travel time represented by the piston flow process (Example 2.1), we obtain

$$\mathrm{E}_z(t^N) = \int_0^\infty t'^N \,\delta\Big(t' - \frac{z\theta}{J_w}\Big)\, dt' = \Big(\frac{z\theta}{J_w}\Big)^N = \Big(\frac{z}{V}\Big)^N \ . \tag{2.54}$$

Thus, the mean travel time to depth z is z/V, and the variance or spread of travel times about the mean is zero by (2.30).

Example 2.4 TRAVEL TIME MOMENTS OF THE CDE

The travel time moments of the CDE are most easily derived using the formula (2.29) relating the Nth moment to the Nth derivative of the Laplace transform of $f^f(z,t)$. Since only ξ in (2.50) depends on s, we may calculate the derivatives of f^f with respect to s by the chain rule of differentiation (Kaplan, 1984). Therefore, the derivatives of $\widehat{f^f}$ (2.50) with respect to s may be written as, using (2.48),

$$\frac{d\widehat{f^f}}{ds} = \frac{d\xi}{ds}\frac{d\widehat{f^f}}{d\xi} = \frac{2D}{V^2\xi}\Big(-\frac{zV}{2D}\Big)\widehat{f^f} = -\frac{z}{V\xi}\widehat{f^f} \tag{2.55}$$

$$\frac{d^2\widehat{f^f}}{ds^2} = \frac{d\xi}{ds}\Big(\frac{z}{V\xi^2} + \frac{z}{V\xi}\frac{zV}{2D}\Big)\widehat{f^f} = \Big(\frac{2Dz}{V^3\xi^3} + \frac{z^2}{V^2\xi^2}\Big)\widehat{f^f} \ . \tag{2.56}$$

Thus, using (2.29), and noting that at $s = 0$, $\xi = 1$ and $\widehat{f^f} = 1$, we find that the travel time mean and variance for the CDE are (Jury and Sposito, 1985; Valocchi, 1985)

$$\mathrm{E}_z(t) = \frac{z}{V} \ , \tag{2.57}$$

$$\mathrm{Var}_z(t) = \frac{2Dz}{V^3} \ . \tag{2.58}$$

Equations (2.57)–(2.58) express two important attributes of the CDE. The mean travel time is the same for all values of the dispersion coefficient D, and it is identical to the mean travel time (2.54) of the piston flow model. This confirms the statement made earlier that in the CDE the dispersive motion within the moving fluid is random and does not contribute to mean displacement. Equation (2.58) states that the variance or spread of travel times about the mean value grows proportionally to the first power of distance from the point ($z = 0$) of injection of the solute pulse. The implications of this behavior will be explored more fully in the next section.

These four examples should serve to illustrate that process-oriented models meeting the assumptions of the transfer function approach can be recast as convolution integrals of the form of (2.4). Furthermore, the travel time pdf of a given process model embodies all of the information about the model, and can effectively replace the differential equation describing the process. Information about the behavior of the model can be obtained by examining the travel time moments. We will find that representing different models in terms of their travel time pdfs will provide an effective means of comparing their behavior in a given application in the soil, and of deciding how to test the assumptions of a particular model in a given experiment. This latter feature should become clearer in the next section.

2.6 Process Assumptions and Travel Time PDFs

Process model assumptions must be verified by experimental observation. However, only certain types of experiments will test the validity of a given model assumption; mere agreement between observed and predicted concentrations may not be sufficient. For example, a curious phenomenon occurred in our history of soil physics: the convection-dispersion equation (2.37) became accepted as the correct model of solute transport through soil without having its assumptions tested adequately in experiments. The experiments that were performed to illustrate the agreement of the model with observation were invariably column outflow experiments (see for example the review by Biggar and Nielsen (1967)), which merely demonstrated that the CDE could be manipulated into matching the shape of the outflow concentrations. However, as long as the solute outflow concentration (assumed to represent the flux concentration at the end of the column) at one specific distance (say $z = \ell$) is the only observation that is being used to test the model, any travel time pdf $f^f(\ell, t)$ with the same shape as the pdf of the CDE (2.51) at $z = \ell$ would predict exactly the same outflow flux concentration as the CDE.

For example, consider the following three parametric models of the travel time flux pdf $f^f(\ell, t)$:[9]

Fickian pdf (CDE)

$$f^f(\ell, t) = \frac{\ell}{2\sqrt{\pi D t^3}} \exp\Big(-\frac{(\ell - Vt)^2}{4Dt}\Big) \tag{2.59}$$

Lognormal pdf

$$f^f(t) = \frac{1}{\sqrt{2\pi}\sigma t} \exp\Big(-\frac{(\ln(t) - \mu)^2}{2\sigma^2}\Big) \tag{2.60}$$

Gamma pdf

$$f^f(t) = \frac{\beta^{1+\alpha} t^\alpha}{\alpha!} \exp(-\beta t) \tag{2.61}$$

Although the three pdfs have quite different functional forms, they have remarkably similar shapes at a particular ℓ when their parameters are optimized to each other (Fig. 2.2).

Therefore, when represented as a transfer function with (2.4), any one of them would do an equally good (or poor) job of modeling the flux concentration data at $z = \ell$ from a column outflow experiment. Equivalently, the normalized nonparametric outflow data (2.18) from a narrow pulse inflow experiment could be used in the transfer function (2.4) to describe a different experiment which had a time-varying solute inflow concentration. No model of the transport process within the column is required to accomplish this.

The relationship between model assumptions and experimental observations is crucial to understand, because one of the major roles of transfer functions will be to design experimental tests of model hypotheses. The relationship between column outflow experiments and the process hypotheses of the CDE (or any other model) might become clearer if we ask two questions about a travel time pdf $f^f(\ell, t)$ which has been measured in an experiment conducted at steady state water flow in a column with a constant water content.

[9] The lognormal and gamma models do not explicitly contain the depth parameter ℓ because they are only required to model a function of time. The Fickian pdf of the CDE contains ℓ explicitly because it is a process model with specific predictions about the z-dependence of $f^f(z, t)$ as well as how it behaves with time at a particular ℓ.

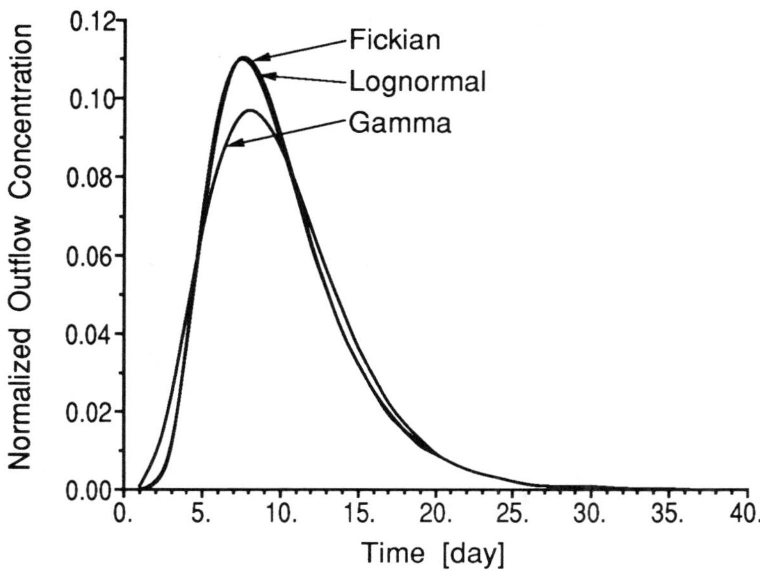

Figure 2.2: *Normalized (unit area) solute outflow concentrations represented by the Fickian, lognormal, and gamma pdfs fitted to each other by the method of moments (see Problems 2.4 and 2.8).*

Question 1 How can a pdf $f^f(\ell, t)$ measured in a soil column of length ℓ at one constant water flow rate be used to predict what the outflow concentration would look like in the same column at a different constant water flow rate?

Question 2 How can a pdf $f^f(\ell, t)$ measured in a soil column of length ℓ at a constant water flow rate be used to predict what the outflow concentration would look like under the same conditions (same soil, water flow rate, etc.), but at a different distance z not equal to ℓ from the inlet end?

In general, we have no way of answering these questions. The outflow concentrations referred to in these questions can be predicted only by extrapolating beyond the experimental observations,[10] which can only be done with the aid of a model of how the system behaves as a function of the variables of the process.

[10]In contrast, the transfer function model (2.4) represents a convolution of the input signal with the experimental observation $f(\ell, t)$.

Let's first look at what is required to answer Question 1. In general, it might be difficult to figure out what the outflow concentration will look like, because when a soil column is subjected to a higher water flow rate, the water content should increase if the column is not saturated. Because the soil has a higher water content, one might expect both the mean travel time and its variance to be different than at a lower water content. Calculation of this new water content could require using a rather complicated water transport model (like the Richards equation describing soil water flow (Baver et al., 1972)). This predicted water content would then have to be combined with a solute transport model that explicitly states the relationship between its parameters and the water content to answer Question 1. For example, the CDE model (2.37) can answer both Question 1 and 2 (but not necessarily correctly!) if its model parameters θ (volumetric water content) and $\lambda := D/V$ (dispersivity) are assumed to be constants which are independent of flow rate and distance along the column (see Problem 2.7). The CDE differential equation (2.37) contains a process description which models the dependence of the concentration on z and t, and provides a description of the relationship between its parameter V and the experimental parameters (J_w and θ). Therefore, once the θ-dependence of the model has been postulated, the CDE can predict how $C^f(z,t)$ will change from the observation defining the pdf $f(z,t)$ at one $z = \ell$, and in the setting addressed in the two questions. However, we can also put additional assumptions into the other model pdfs to allow them to perform the extrapolations posed in the questions.

We can allow each model to provide the information for answering Question 1 simply by assuming that any change in the solute transport volume geometry of the soil column will be negligible as the flux is changed. Here is a statement of this hypothesis:

> **Hypothesis 1** In the soil in Question 1, the water content does not change as the water flux J_w is changed to a new constant value, and solute outflow at any water flux rate will look exactly the same if it is plotted as a function of cumulative drainage $I := J_w t$.

This hypothesis actually has been shown to be reasonably accurate for repacked soil columns in the laboratory (Khan, 1988), and also for area-averaged flux concentrations monitored by solution samplers at various depths on a large field where solute was leached by sprinkler irrigation and rainfall (Jury et al., 1990). It has also been tested using computer simulation with the Richards equation for water flow (Wierenga, 1977).

With the aid of Hypothesis 1, we have a model that will answer Question 1. Assume that we measure a travel time pdf $f_1(t)$ in a soil column of length ℓ

under conditions of steady water flux J_1 and we want to predict what the travel time pdf $f_2(t)$ will be when the same column is irrigated at a steady state water flux J_2. Since both experimental outflows would look the same as a function of $I = J_w t$, it can be shown (see Problem 2.5) that the cdf and pdf which obey Hypothesis 1 are given by

$$P_2(t) = P_1\Big(\frac{J_2}{J_1}t\Big) . \tag{2.62}$$

$$f_2(t) = \frac{J_2}{J_1} f_1\Big(\frac{J_2}{J_1}t\Big) . \tag{2.63}$$

With a little modification, we can transform each of our pdfs (2.59)–(2.61) so that they obey (2.63). The Fickian pdf (2.59) representing the CDE model will obey (2.63) if we let $V = J_w/\theta$, where θ is a constant and if $D = \lambda V$, where λ is a constant (called the dispersivity). Thus a CDE with a constant water content and a dispersivity obeys (2.63).

The lognormal pdf model (2.60) will obey (2.63) if we make its parameter μ a function of J_w by the relation $\mu := \mu_0 - \ln(J_w)$[11], and the gamma pdf (2.61) will obey (2.63) if $\beta := \beta_0 J_w$, where μ_0 and β_0 are constants. With these modifications, all three pdfs would make the same predictions in answering Question 1. Consequently, data from a series of outflow experiments at different flow rates in the same column is not sufficient to validate the assumptions of the CDE. Rather, it would be useful for testing the validity of Hypothesis 1.

Question 2 is a more interesting one, because it is asking about behavior inside of or beyond the end of the solute transport volume, neither of which can be deduced from an observation at one distance without knowing the physics of the transport process. One very useful method of deducing the way in which solute concentrations behave with depth is to look at the travel time moments at different distances from the inlet end.

Let's begin by examining what the CDE predicts about the solute transport process for an arbitrary location z, for constant J_w, θ, V and D. As shown in Example 2.4 above, the mean and variance of the travel time pdf of the CDE model are given by (2.57)–(2.58). Hence, in a homogeneous soil under steady state water flow, both the mean travel time and its variance increase linearly with distance from the point of solute input for a solute transport process that obeys the CDE. Thus, the CDE with constant θ, λ makes a specific prediction about the behavior of the solute concentration at different z.

---

[11]Strictly, this should be written as $\ln(J_w/J_{w_0})$, where J_{w_0} is a unit flux in the desired system of units.

The other two pdfs (2.60) and (2.61) don't contain z, so they aren't able to answer Question 2 (yet). As an example of a way to relate travel time pdfs at different z in a manner which will give rise to behavior which is very different from the CDE, consider the following hypothesis (Jury, 1982):

> **Hypothesis 2** In a transport volume which has homogeneous average properties along its length[12] the probability that a solute molecule added at $t = 0$ at the inlet end $z = 0$ will arrive at a depth z in a time t_0 less than or equal to t is the same as the probability that it will arrive at a depth ℓ in a time t_0 less than or equal to tL/z.

This statement of probability for travel times physically corresponds to a transport volume in which the faster and slower solute molecules move in a manner which isolates them from their neighbors, as if they were trapped in stream tubes that couldn't communicate with each other. Thus, a model obeying this hypothesis is diametrically opposite to the CDE, which assumes that lateral mixing of solute between regions of different water flow occurs over a time period which is much shorter than the time required for solute to be moved from the inlet end to the outlet end of a transport volume (Taylor, 1953). Mathematically, the hypothesis can be restated in terms of the travel time cdf $P(t)$ as

$$P(z,t) = P\left(\ell, \frac{t\ell}{z}\right) , \tag{2.64}$$

or, using (2.24),

$$f^f(z,t) = \frac{\ell}{z}\, f^f\left(\ell, \frac{t\ell}{z}\right) . \tag{2.65}$$

Equation (2.65) provides a model for answering the second question. The pdf $f^f(\ell, t)$ at the depth of observation is transformed into a predicted pdf for $f^f(z,t)$ by multiplying it by ℓ/z and substituting $t\ell/z$ in place of t. It is easy to show (Problem 2.10) that the Fickian pdf (2.59) does not obey (2.65).

The travel time moments of a pdf $f^f(z,t)$ which obeys (2.65) have some interesting properties. The Nth moment (2.28) is equal to

$$\begin{aligned} \mathrm{E}_z(t^N) &= \int_0^\infty t^N f^f(z,t)\, dt = \int_0^\infty t^N \frac{\ell}{z}\, f^f\left(\ell, \frac{t\ell}{z}\right) dt \\ &= \left(\frac{z}{\ell}\right)^N \int_0^\infty y^N f^f(\ell, y)\, dy = \left(\frac{z}{\ell}\right)^N \mathrm{E}_\ell(t^N) , \end{aligned} \tag{2.66}$$

[12]Any properties which vary locally belong to the same distribution.

where $\mathrm{E}_\ell(t^N)$ is the Nth travel time moment evaluated at the reference outflow depth $z = \ell$. Thus, the mean and variance of the travel time distribution of a soil obeying Hypothesis 2 are

$$\mathrm{E}_z(t) = \frac{z}{\ell}\ \mathrm{E}_\ell(t)\ , \tag{2.67}$$

$$\mathrm{Var}_z(t) = \left(\frac{z}{\ell}\right)^2\ \mathrm{Var}_\ell(t)\ . \tag{2.68}$$

Comparing (2.58) and (2.68), we see a real difference in the way that solute moves according to the two models. The new model, which is based on Hypothesis 2, predicts that solute travel times will spread out (as defined by the variance) at a rate proportional to the square of the distance from the inlet end, whereas the CDE model predicts that the rate of spreading should only be proportional to the first power of distance. If we were to test the CDE model on a system that behaved according to Hypothesis 2, it might be manipulated to agree perfectly with the outflow concentration record at one depth, but it would appear as though its dispersion coefficient D was increasing linearly with distance if it was fitted repeatedly to a set of different outflow curves at different distances from the source of solute input. This phenomenon, which has been observed many times in ground water experiments (see review by Gelhar et al., 1985), and recently in a field soil (Butters and Jury, 1989), is sometimes called the dispersion scale effect (Fried, 1975).

Clearly, it is the behavior of the travel time variance with distance that distinguishes the CDE from other models. Therefore, experimental tests of the validity of the CDE in soil columns should have monitored the outflow concentration as a function of time at various distances from the inlet end (or as a function of distance at various times, as will become clear in Chapter 3). However, except for a very recent study (Khan and Jury, 1990), this test of the model has not been performed in the laboratory. Instead, the CDE has been used to describe outflow concentrations only at one distance from the inlet end, where it could not even be distinguished from a model obeying Hypothesis 2.

Each of the travel time pdfs (2.59)–(2.61) can be modified easily so that they obey (2.65) at different distances from the inlet end of the column in a single steady state flow experiment, by defining depth-dependent parameters through the following relations

$$\text{Fickian} \qquad V_z = V_\ell \quad \text{and} \quad D_z = \frac{z}{\ell} D_\ell\ , \tag{2.69}$$

$$\text{Lognormal} \qquad \mu_z = \mu_\ell + \ln(z/\ell) \quad \text{and} \quad \sigma_z = \sigma_\ell\ , \tag{2.70}$$

$$\text{Gamma} \qquad \alpha_z = \alpha_\ell \quad \text{and} \quad \beta_z = \frac{\ell}{z}\,\beta_\ell \ , \tag{2.71}$$

where now the parameters with the subscript ℓ are constant for all z and the depth ℓ has become a calibration depth for defining each model.

2.7 Stochastic-Convective and Convective-Dispersive Models

We will call any process that obeys (2.58) (i.e. whose travel time variance grows linearly with distance) a *convective-dispersive* process, and one that obeys (2.68) (quadratic growth of variance with distance) a *stochastic-convective* process (Simmons, 1982). For example, the Fickian pdf (2.51) with parameters V_z and D_z which obey (2.69) is now stochastic-convective, and no longer is a solution of the CDE differential Equation (2.37) (Simmons, 1986; see also Problem 2.2).

The lognormal stochastic convective model ((2.60) with parameters defined by (2.70)), first proposed by Jury (1982), has a special name, the convective lognormal transfer function model (CLT) (Jury, 1987).

$$f^f(z,t) = \frac{1}{\sqrt{2\pi}\,\sigma_\ell t}\,\exp\Bigl(-\frac{\bigl(\ln(t\ell/z) - \mu_\ell\bigr)^2}{2\sigma_\ell^2}\Bigr)\ . \tag{2.72}$$

The pdfs of the CDE (2.59) and CLT (2.72) are compared in Figure 2.3 below.

The three graphs in Figure 2.3 show some of the interesting features of the convective lognormal CLT model which obeys (2.65). It is virtually indistinguishable from the CDE model when the parameters of each model are optimized to each other or to a common data set at a given depth (as in the top figure at 50 cm). After calibration, however, the CLT will spread out much faster than the CDE, and will have about the same average travel time to any depth of monitoring.

This chapter has been confined to the representation of solute flux concentrations at the outlet of a solute transport volume. For such a task, the travel time pdf (or, more generally, the solute lifetime distribution) can be used in place of a process model which is required to characterize the transport and transformation process inside of the volume. In the next chapter, the discussion will be extended to include characterization of the solute resident concentrations, which describe the solute mass density at different locations in space.

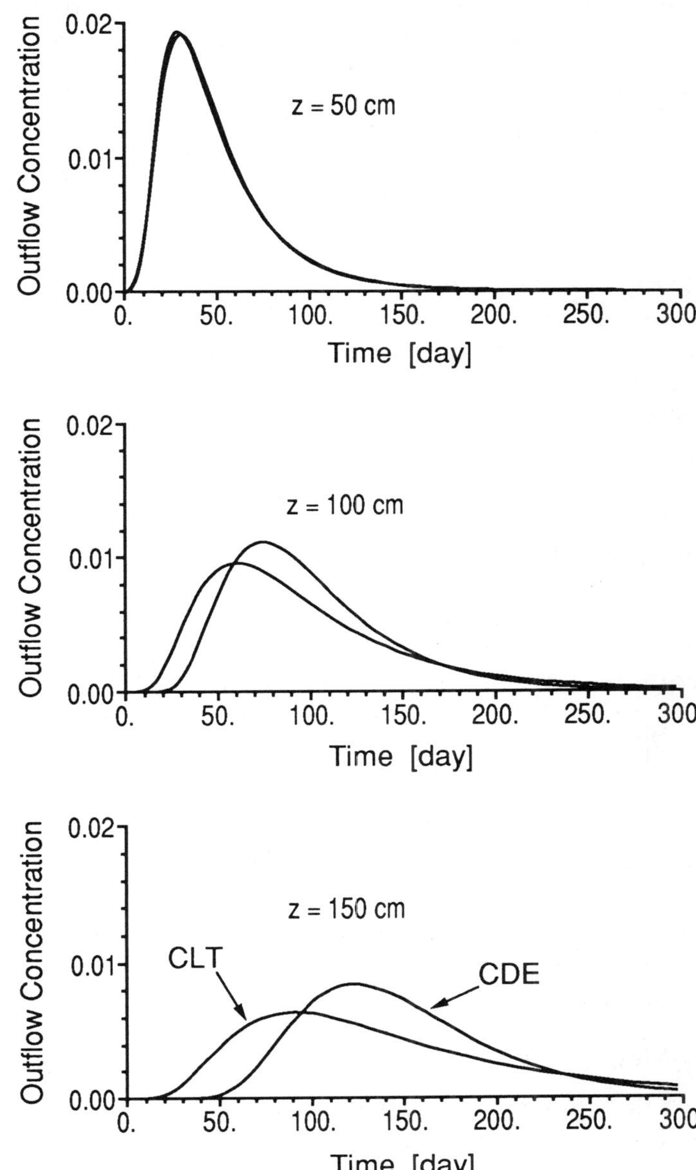

Figure 2.3: *Normalized solute outflow concentrations predicted at three depths by the CDE and CLT models after common calibration at ℓ = 50 cm by the method of moments.*

Problems

Problem 2.1 Derive the inverse (2.51) of the Laplace transform of the travel time pdf for the CDE (2.50) using the table of transforms.

Problem 2.2 Show that the stochastic-convective Fickian pdf ((2.51) with (2.69)) is not a solution of the CDE (2.37).

Problem 2.3 Calculate the Laplace transform of the flux concentration for the following three input boundary conditions

$$C^f(0,t) = \delta(t) \; , \tag{2.73}$$

$$C^f(0,t) = H(t) \; , \tag{2.74}$$

$$C^f(0,t) = H(t) - H(t - \Delta t) \; . \tag{2.75}$$

Problem 2.4 If $f(t)$ is a lognormal travel time pdf

$$f(t) = \frac{1}{\sqrt{2\pi}\,\sigma t} \exp\Big(-\frac{(\ln(t)-\mu)^2}{2\sigma^2}\Big) \; , \tag{2.76}$$

prove that the Nth travel time moment is given by

$$\mathrm{E}(t^N) = \int_0^\infty t^N f(t)\,dt = \exp(N\mu + N^2\sigma^2/2) \; . \tag{2.77}$$

Problem 2.5 Prove the relation given in (2.63).

Problem 2.6 Calculate the flux concentration of the CDE corresponding to a step function input of solute

$$C^f(0,t) = C_0\, H(t) \; , \tag{2.78}$$

where $H(t)$ is the Heaviside function.

Problem 2.7 Prove that all steady state solutions of the CDE for the flux pdf will be invariant when plotted against cumulative drainage $I = J_w t$ if θ and $\lambda = D/V$ are constant.

Problem 2.8 Calculate the travel time moments of the gamma pdf (2.61).

Problem 2.9 Using the definition of the Laplace transform (A.1), show that (2.29) is valid.

Problem 2.10 Show that the Fickian pdfs (2.51) and (2.59) do not obey the stochastic-convective hypothesis (2.65).

Chapter 3

Flux and Resident Concentrations

3.1 Types of Solute Concentration

It is important to realize that there are different types of solute concentration that appear in various models, or that are measured by different devices. In the travel time formulation we have been developing thus far, the inflow and outflow concentrations that move through the fixed planes forming the boundaries of the transport volume are called flux concentrations. A *flux concentration* C^f is defined for one dimensional flow as the ratio of the solute mass flux to the water flux. (The higher dimensional definition is slightly more complicated; see discussion by Sposito and Barry, 1987). To picture a flux concentration, imagine that there is an invisible plane perpendicular to the solute flux direction, and that the mass of solute crossing the plane during a short time interval is divided by the volume of water that crosses the plane over the same period. This mass/volume is a flux concentration. Therefore, we can write

$$J_s = C^f J_w \ , \tag{3.1}$$

where s and w refer to the solute and water fluxes respectively. A flux concentration is thus a flow-weighted concentration, and would be measured by a perfectly passive device in the soil that allowed solution to enter it at exactly the same rate as the flow rate in the absence of the device.

Another type of concentration is the instantaneous mass density of solute in a fixed volume of space. The total *resident concentration* C_t^r is defined as the mass of solute per unit volume of soil. Therefore, total resident concentrations could be measured by soil coring, in situ electrical resistance measurements, or other probes of the system that "freeze" the positions of the solute molecules and record their density.

The total resident concentration may be composed of solutes present in several phases in the soil (e.g. dissolved in solution, gaseous, adsorbed, immiscible liquid, etc.). When more than one phase is mobile in the soil, the flux concentration as defined by (3.1) would consist of more than one phase. To avoid confusion in that case, the solute mass flux form of the transfer function (2.1) should be used. For most of this book, the discussion of solute transport will be confined to nonvolatile solutes that are either dissolved in solution or adsorbed to the soil solid phase. Therefore, the solute flux concentration will be associated with dissolved solutes only, unless otherwise stated.

The total resident concentration can only be divided up into its phase constituents in the soil with the aid of process models relating the phases. However, when only the flux concentration of the dissolved phase leaves the transport volume, it may be related by the principle of mass balance to the total resident concentration, as shown below.

3.2 The Solute Conservation Equation

The solute conservation equation for a compound which does not lose or gain mass by reactions within the transport volume may be written for one dimensional flow as

$$\frac{\partial C_t^r}{\partial t} + J_w \frac{\partial C^f}{\partial z} = 0 . \tag{3.2}$$

This relationship follows directly from (3.1), and the definitions of the solute flux and the resident concentration. For a general solute transport process in semi-infinite soil ($0 < z < \infty$), (3.2) requires an initial condition and an inlet boundary condition.[1]

$$C_t^r(z,0) = C_0^r(z) \tag{3.3}$$

$$C^f(0,t) = C_0^f(t) \tag{3.4}$$

Equation (3.2) may be integrated formally to yield expressions relating the two concentrations (assuming constant J_w). When the flux concentration is known, the resident concentration is calculated by integrating (3.2) over t

$$C_t^r(z,t) = C_0^r(z) - J_w \frac{\partial}{\partial z} \int_0^t C^f(z,t')\, dt' . \tag{3.5}$$

[1]The condition at $z = \infty$ is assumed to remain unchanged from its initial state.

Conversely, when the resident profile is known, the flux concentration may be obtained be integrating (3.2) over z

$$C^f(z,t) = C_0^f(t) - \frac{1}{J_w}\frac{\partial}{\partial t}\int_0^z C_t^r(z',t)\,dz' \tag{3.6}$$

These relations become particularly simple if the Laplace transform operation is applied to both sides

$$\widehat{C}_t^r(z;s) = \frac{C_0^r(z)}{s} - \frac{J_w}{s}\frac{d\widehat{C}^f(z;s)}{dz} \tag{3.7}$$

$$\widehat{C}^f(z;s) = \widehat{C}_0^f(s) - \frac{s}{J_w}\int_0^z \widehat{C}_t^r(z';s)\,dz' + \frac{1}{J_w}\int_0^z C_t^r(z',0)\,dz' \,. \tag{3.8}$$

Equations (3.5)–(3.6) are just an alternate statement of solute mass balance. In (3.5), the resident concentration may be calculated only if the space and time dependence of the flux concentration is known. As discussed in Chapter 2, this information requires either a model linking the flux pdfs at different distances from the inlet end of the transport volume, or a series of experiments measuring $C^f(z,t)$ at different depths. If data is used, the outflow distances should be close enough together to allow the integral in (3.5) to be evaluated accurately. Similarly, a flux concentration could not be calculated with (3.6) from a single resident concentration profile, but would require a model or repeated observation of the time evolution of that profile. Thus, since experimental information that is comprehensive enough to meet these requirements is never available, the principal value of these equations lies in linking the flux and resident concentrations of transport process models. The next example will illustrate how these expressions may be used in that capacity.

Example 3.1 RESIDENT CONCENTRATIONS OF THE CDE

We wish to calculate the total resident concentration C_t^r corresponding to the entry of a narrow pulse (delta function) of solute flux concentration through the inlet end $z = 0$ at $t = 0$, assuming that there is no solute present in the system initially ($C_t^r(z,0) = 0$). The flux concentration C_f for this problem is the Fickian pdf (2.51). The Laplace transform $\widehat{C}^f$ of the Fickian pdf is given by (see Example 2.2)

$$\widehat{C}^f(z;s) = \exp\left(\frac{Vz}{2D}(1-\xi)\right) \,, \tag{3.9}$$

where

$$\xi = \sqrt{1 + 4sD/V^2} \,. \tag{3.10}$$

Thus, the Laplace transform of the total resident concentration is, using (3.7)

$$\begin{aligned} \widehat{C}_t^r(z;s) &= -\frac{J_w}{s}\frac{d\widehat{C}^f(z;s)}{dz} = -\frac{\theta V^2(1-\xi)}{2Ds}\exp\left(\frac{Vz}{2D}(1-\xi)\right) \\ &= \frac{2\theta}{1+\xi}\exp\left(\frac{Vz}{2D}(1-\xi)\right) . \end{aligned} \tag{3.11}$$

The inverse transform of (3.11) is derived in Problem 3.1 using the Table of Laplace Transforms (Appendix C). It is equal to

$$C_t^r(z,t) = \frac{\theta V}{\sqrt{\pi D t}}\exp\left(-\frac{(z-Vt)^2}{4Dt}\right) - \frac{\theta V^2}{2D}\exp\left(\frac{Vz}{D}\right)\operatorname{erfc}\left(\frac{z+Vt}{\sqrt{4Dt}}\right) , \tag{3.12}$$

where $\operatorname{erfc}(x) := 1 - \operatorname{erf}(x)$ is the complementary error function, and the error function $\operatorname{erf}(x)$ is defined by (Abramowitz and Stegun, 1970)

$$\operatorname{erf}(x) := \frac{2}{\sqrt{\pi}}\int_0^x \exp(-y^2)\,dy . \tag{3.13}$$

Figure 3.1 shows the CDE flux concentration (2.51) at various distances from the inlet end, and the corresponding resident concentration (3.12) at various times.
The solution for the resident CDE when the inlet condition is the narrow square pulse (2.15) was derived by Lindstrom et al. (1967) and Gershon and Nir (1969). It can be shown to reduce to (3.12) in the limit as $\Delta t \to 0$ and $C_0\Delta t \to 1$.

By comparing the flux (3.1) and resident (2.34) concentration forms of the solute flux for the one dimensional CDE, we can relate the flux and resident concentrations directly (Parker and Van Genuchten, 1984)

$$C^f = C_l^r - \frac{D}{V}\frac{\partial C_l^r}{\partial z} , \tag{3.14}$$

where C_l^r is the solute mass per solute volume resident in the dissolved phase. Equation (3.14) will be valid whenever the solute flux is described by (2.34).

The total resident concentration can be broken up into component parts by making various model assumptions. For example, in soils which have a natural structure, not all of the wetted pore space is active in mass transport of solute; part of the water phase is stagnant and can only be reached by solutes diffusing out of the part of the solution phase which is flowing. In this case, one may wish to divide the total resident concentration (assuming no adsorption) into the sum of a mobile and a stagnant or "immobile" liquid phase (Coats and Smith, 1956; Van Genuchten and Wierenga, 1976)

$$C_t^r = \theta_m C_m^r + \theta_{im} C_{im}^r , \tag{3.15}$$

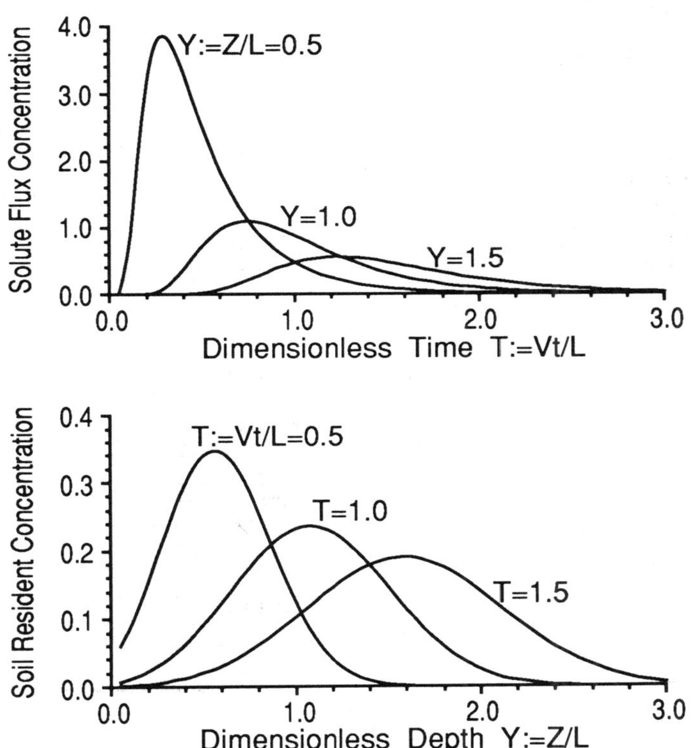

Figure 3.1: *Solute flux concentrations (top) and the corresponding total resident concentrations (bottom) for a soil with a Peclet number* $P := LV/D = 10$ *and* $\theta = 0.25$.

where the subscripts m and im refer to mobile and immobile, respectively. This model is examined in the next example.

Example 3.2 THE MOBILE-IMMOBILE WATER MODEL (MIM)

The differential equations describing the two region, mobile-immobile water model for a non-adsorbing solute which does not decay under steady state water flow in homogeneous soil may be written as (Van Genuchten and Wierenga, 1976)

$$\theta_m \frac{\partial C_m^r}{\partial t} + \theta_{im} \frac{\partial C_{im}^r}{\partial t} = \theta_m D_m \frac{\partial^2 C_m^r}{\partial z^2} - J_w \frac{\partial C_m^r}{\partial z} \tag{3.16}$$

$$\theta_{im} \frac{\partial C_{im}^r}{\partial t} = \alpha (C_m^r - C_{im}^r) , \tag{3.17}$$

where J_w is the steady water flux rate, D_m is the diffusion-dispersion coefficient in the mobile zone, and α is a rate parameter for exchange between the phases. Note that both C_m^r and C_{im}^r in (3.16)–(3.17) are resident concentrations. We wish to calculate the Laplace transform of the solution for the mobile water flux concentration C_m^f corresponding to a narrow pulse input of solute flux concentration when the medium is infinite in the z-direction and initially devoid of solute, i.e.

$$C_m^r(z,0) = 0 , \quad C_{im}^r(z,0) = 0 , \quad C_m^f(\infty,t) = 0 , \quad C_m^f(0,t) = \delta(t) . \tag{3.18}$$

After Laplace transformation as in Example 2.2, (3.16) and (3.17) may be written as

$$s\theta_m \widehat{C}_m^r + s\theta_{im} \widehat{C}_{im}^r = \theta_m D_m \frac{d^2 \widehat{C}_m^r}{dz^2} - J_w \frac{d\widehat{C}_m^r}{dz} \tag{3.19}$$

$$s\theta_{im} \widehat{C}_{im}^r = \alpha(\widehat{C}_m^r - \widehat{C}_{im}^r) . \tag{3.20}$$

When $\widehat{C}_{im}^r$ is solved for in (3.20) and inserted into (3.19), the result may be written

$$\theta_m D_m \frac{d^2 \widehat{C}_m^r}{dz^2} - J_w \frac{d\widehat{C}_m^r}{dz} - g(s)\theta_m \widehat{C}_m^r = 0 , \tag{3.21}$$

where

$$g(s) = s + \frac{s\alpha\theta_{im}/\theta_m}{\alpha + s\theta_{im}} . \tag{3.22}$$

The mobile flux concentration $\widehat{C}_m^f$ is related to the mobile resident concentration through the solute flux equation (3.1), and (2.34) with $C_l^r \to C_m^r$,

$$\widehat{C}_m^f \equiv \widehat{C}^f = \frac{\widehat{J}_s}{J_w} = \widehat{C}_m^r - \frac{D_m}{V_m} \frac{d\widehat{C}_m^r}{dz} , \tag{3.23}$$

where $V_m := J_w/\theta_m$ is the mobile solute velocity. Note that the mobile flux concentration (3.23) is identical to the total flux concentration, since solute in the immobile phase cannot exit the transport volume except through the mobile phase. Since $d\widehat{C}_m^r/dz$ also satisfies (3.21), it follows from (3.23) that

$$\theta_m D_m \frac{d^2 \widehat{C}^f}{dz^2} - J_w \frac{d\widehat{C}^f}{dz} - g(s)\theta_m \widehat{C}^f = 0 . \tag{3.24}$$

Thus, the flux concentration satisfies the same differential equation as the mobile resident concentration.

Equation (3.24) is solved in the standard manner by a trial substitution $\widehat{C}^f = \exp(mz)$ (see (2.45)) which produces the two roots m of

$$m = \frac{V_m}{2D_m} \pm \frac{V_m}{2D_m}\sqrt{1 + \frac{4g(s)D_m}{V_m^2}} = \frac{V_m}{2D_m}(1 \pm \zeta) \, , \tag{3.25}$$

where

$$\zeta := \sqrt{1 + \frac{4g(s)D_m}{V_m^2}} \, . \tag{3.26}$$

After applying the upper and lower boundary conditions (3.18) for $\widehat{C}^f$ to the general solution, we obtain

$$\widehat{C}^f = \widehat{f}^f(z;s) = \exp\Big(\frac{V_m z}{2D_m}(1-\zeta)\Big) \, . \tag{3.27}$$

Since we have added a $\delta(t)$-pulse of solute flux concentration to the inlet end, (3.27) is the Laplace transform of the travel time pdf $f^f(z,t)$ for this model.

The total resident concentration, which is obtained from (3.27) using (3.7), may be written as

$$\begin{aligned} \widehat{C}_t^r &:= \frac{1}{s}\frac{V_m J_w}{2D_m}(\zeta - 1) \exp\Big(\frac{V_m z}{2D_m}(1-\zeta)\Big) \\ &= \frac{1}{s}\frac{2\theta_m g(s)}{1+\zeta} \exp\Big(\frac{V_m z}{2D_m}(1-\zeta)\Big) \, . \end{aligned} \tag{3.28}$$

The flux and total resident concentrations of the MIM model calculated by numerical inversion of (3.27)–(3.28) (see Appendix B) are plotted in Figures 3.2–3.3, for three values of the dimensionless rate parameter $W := \alpha\ell/J_w$.

Some attributes of the MIM model may be discovered by comparing Figures 3.2–3.3 with the corresponding $Y = 1$ and $T = 1$ curves of Figure 3.1 for the CDE. First, because only part of the wetted pore space is mobile, the peak solute flux concentration arrives well ahead of the piston flow breakthrough time $\ell\theta/J_w$. Second, the rate limited exchange of solute between the mobile and stagnant zones produces a long solute outflow tail in the flux pdf (i.e. very long travel times), and a correspondingly large quantity of solute which remains near the surface compared to the CDE. These observations may be quantified by examining the travel time moments of the MIM, as shown in the next example.

Example 3.3 TRAVEL TIME MOMENTS OF THE MIM

The travel time moments are most easily obtained as in Example 2.4 by using the formula (2.29) to operate on the Laplace transform (3.27) of the flux pdf. The chain rule of differentiation may be used to differentiate (3.27) with respect to s,

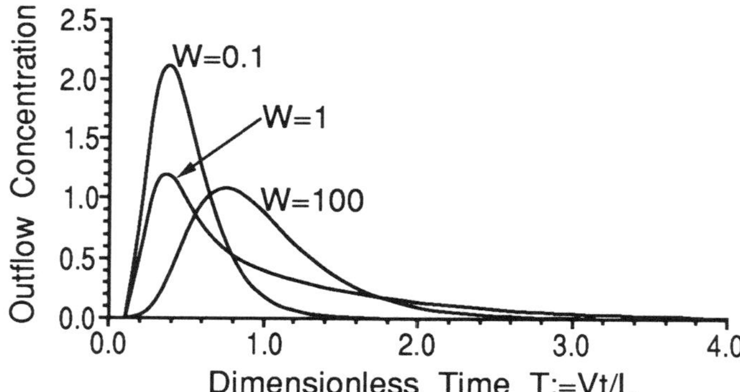

Figure 3.2: *Flux concentration pdfs of the mobile-immobile water model as a function of the rate parameter $W := \alpha\ell/J_w$ plotted at depth $z = \ell$. (Curves plotted by numerical inversion of (3.27)—see Appendix B).*

$$\frac{d\widehat{f}^f}{ds} = \left(\frac{dg}{ds}\right)\left(\frac{d\zeta}{dg}\right)\left(\frac{d\widehat{f}^f}{d\zeta}\right) , \tag{3.29}$$

where g and ζ are defined in (3.22) and (3.26).

These derivatives are

$$\begin{aligned}
\frac{dg}{ds} &= 1 + \frac{\theta_{im}\alpha/\theta_m}{\alpha + s\theta_{im}} - \frac{s\theta_{im}{}^2\alpha/\theta_m}{(\alpha+s\theta_{im})^2} \quad \xrightarrow{s\to 0} \quad \frac{\theta}{\theta_m} ;\\
\frac{d\zeta}{dg} &= \frac{2D_m}{V_m^2\zeta} \quad \xrightarrow{s\to 0} \quad \frac{2D_m}{V_m^2} ;\\
\frac{d\widehat{f}^f}{d\zeta} &= -\frac{zV_m}{2D_m}\widehat{f}^f \quad \xrightarrow{s\to 0} \quad -\frac{zV_m}{2D_m} .
\end{aligned}$$

Thus, the first moment is equal to

$$\mathrm{E}_z(t) = -\frac{d\widehat{f}^f}{ds}\bigg|_{s=0} = \frac{dg}{ds}\frac{z}{V_m\zeta}\widehat{f}^f\bigg|_{s=0} = \frac{\theta z}{\theta_m V_m} = \frac{\theta z}{J_w} , \tag{3.30}$$

which is the same as the mean travel time of the CDE (2.57) and of the piston flow model (2.54). Thus, the mean travel time is unaffected by the presence of immobile water, even though the peak may arrive very early. The long tail compensates for the early arrival. The second derivative may be decomposed as follows, from (3.29)

$$\frac{d^2\widehat{f}^f}{ds^2} = -\left[\frac{d^2g}{ds^2}\frac{z}{\zeta V_m} - \left(\frac{dg}{ds}\right)^2\frac{2D_m z}{V_m^3\zeta^3} - \left(\frac{dg}{ds}\right)^2\frac{z^2}{V_m^2\zeta^2}\right]\widehat{f}^f , \tag{3.31}$$

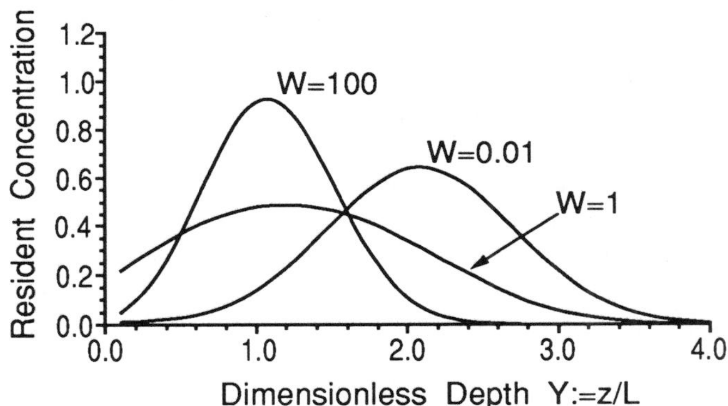

Figure 3.3: *Total resident concentration of the mobile-immobile water model as a function of the rate parameter $W := \alpha\ell/J_w$, plotted at time $t = \ell\theta/J_w$. (Curves plotted by numerical inversion of (3.28)—see Appendix B).*

where

$$\frac{d^2g}{ds^2} = \frac{2s{\theta_{im}}^3\alpha/\theta_m}{(\alpha+s\theta_{im})^3} - \frac{2{\theta_{im}}^2\alpha/\theta_m}{(\alpha+s\theta_{im})^2} \xrightarrow{s\to 0} -\frac{2\theta_{im}^2}{\alpha\theta_m}\,. \tag{3.32}$$

Thus, the second moment is, by (2.29),

$$\mathrm{E}_z(t^2) = \left.\frac{d^2\widehat{f}^f}{ds^2}\right|_{s=0} = \left(\frac{\theta z}{J_w}\right)^2 + \frac{2D_m\theta^2 z}{V_m^3{\theta_m}^2} + \frac{2{\theta_{im}}^2 z}{\alpha J_w}\,. \tag{3.33}$$

Therefore, by (2.30), the variance is

$$\mathrm{Var}_z(t) = \frac{\theta_m}{\theta}\frac{2D_m z}{V^3} + \frac{\theta_{im}}{\theta}\frac{2\theta_{im} z}{\alpha V}\,, \tag{3.34}$$

where $V = J_w/\theta$. Thus, the travel time variance of the MIM is the sum of the dispersive variance within the mobile zone, and a second term representing solute spreading from rate limited diffusive mass transfer between the mobile and immobile zones (Valocchi, 1985).

The MIM model has been compared to observation in a few laboratory tests (Van Genuchten and Wierenga, 1977; Nkedi Kizza et al., 1983; De Smedt and Wierenga, 1984; Schulin et al., 1987). In most cases, the rate parameter and mobile water fraction can only be obtained by simultaneous parameter fitting. However, Rao et al. (1980) were able to establish a relationship between aggregate size and rate parameter for a porous medium containing a single size of spherical aggregates.

3.3 Flux and Resident PDFs

The impulse response function for a narrow pulse (delta function) flux input of solute $f^f(z,t)$ has been interpreted as a travel time pdf giving the distribution of travel times that a solute may experience in moving from 0 to z. In a similar manner, we may define a second impulse response function to the narrow pulse input, namely the solute resident concentration as a function of z at a fixed time t. This resident concentration $f^r(z,t)$ may be interpreted as a travel distance pdf giving the distribution of depths that a solute might reach by the end of the fixed time t (Simmons, 1986a).

The relation between the two pdfs can be derived as follows. If a narrow pulse of solute is added to the inlet end of the transport volume at $t = 0$, then at time t the fraction of the pulse that has passed z must now be found between z and ∞. Therefore, we may write

$$\int_0^t f^f(z,t')\,dt' = \int_z^\infty (f^r(z',t) - f^r(z',0))\,dz' \ . \tag{3.35}$$

If we differentiate both sides of (3.35) with respect to z and t, we obtain the differential form

$$\frac{\partial f^r(z,t)}{\partial t} + \frac{\partial f^f(z,t)}{\partial z} = 0 \tag{3.36}$$

(Simmons, 1986; Dagan and Nguyen, 1989). Thus, if the travel time pdf $f^f(z,t)$ has been calculated with a process model, or measured as a function of z, the corresponding resident concentration pdf may be derived—assuming no resident solute concentration at $t = 0$—from

$$f^r(z,t) = -\frac{\partial}{\partial z}\int_0^t f^f(z,t')\,dt' \ . \tag{3.37}$$

Equation (3.37) may also be Laplace-transformed, resulting in

$$\hat{f}^r(z;s) = -\frac{1}{s}\frac{d\hat{f}^f(z;s)}{dz} \ . \tag{3.38}$$

This resident pdf, because it has been defined with respect to the travel time pdf, therefore applies specifically to the case of a narrow pulse flux input through the surface, which is maintained as a zero flux boundary for $t > 0$.

3.4 Depth Moments of the Resident PDF

Once the resident pdf has been obtained for a given model, its depth moments may be calculated from

$$Z_N(t) := \mathrm{E}_t(z^N) = \int_0^\infty z^N f^r(z,t)\,dz \ . \tag{3.39}$$

Equation (3.39) will be useful in estimating the parameters of models from observations of soil resident concentrations, and also in showing the differences in the perception of processes viewed by resident concentrations (soil sampling) and by flux concentrations (solution sampling). This will become apparent when we discuss the differences between the stochastic-convective and convective-dispersive processes. First, however, we will derive the depth moments of the CDE.

Example 3.4 DEPTH MOMENTS OF THE CDE

The Laplace transform of (3.39) is

$$\widehat{Z}_N = \int_0^\infty z^N \, \widehat{f}^r(z,t)\, dz \; . \tag{3.40}$$

The Laplace transform of the resident pdf $\widehat{f}^r(z,t)$ is given by $\widehat{C}_t^r / J_w$, where $\widehat{C}_t^r$ is the transformed total resident concentration (3.11) corresponding to the narrow pulse input to the CDE.[2] Thus

$$\widehat{f}^r(z,t) = \frac{2}{V(1+\xi)} \exp\left(\frac{Vz}{2D}(1-\xi)\right) , \tag{3.41}$$

where ξ is defined as

$$\xi := \sqrt{1 + \frac{4sD}{V^2}} \; . \tag{3.42}$$

Therefore, the integration in (3.40) may be carried out analytically (see (D.39)), with the result

$$\widehat{Z}_N = \frac{2N!}{V(1+\xi)} \left(\frac{2D}{V(\xi-1)}\right)^{N+1} \; . \tag{3.43}$$

Equation (3.43) may be inverted analytically for the first few depth moments (see Problem 3.3). Figure 3.4 shows a plot of the effective resident solute velocity $V^r_{\mathit{eff}} := dZ_1/dt$ of the pulse in space and the effective dispersion coefficient[3] $D^r_{\mathit{eff}} := 0.5\, d\mathrm{Var}/dt$ describing the spreading of the pulse, where $\mathrm{Var} = Z_2 - Z_1^2$ is the spatial variance of the solute about its center of mass position in the soil at time t.

At first, Figure 3.4 may seem confusing. The reason that the velocity and dispersion coefficients of the CDE model seem to be changing with time when analysed by depth moments is because of the surface boundary condition, which does not allow solute to diffuse upward. Because this breaks the symmetry of the diffusion process, the center of mass of the solute pulse moves downward faster and spreads out slower than it would in an infinite medium. Only when the solute is far below the surface does the influence of the upper boundary disappear.

---

[2]This should be clear by examining (3.2) and (3.36), applied to the case where $C^f = f^f(z,t)$.

[3]This definition of the dispersion is developed in Chapter 7. See also Simmons (1986b).

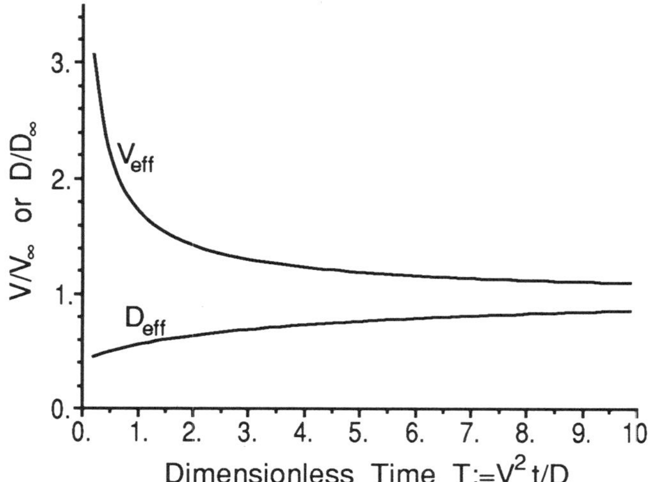

Figure 3.4: *Effective velocity and dispersion coefficients calculated from the depth moments of the CDE, plotted as a ratio of the final values.*

3.5 Resident PDF of the Stochastic-Convective Model

A stochastic-convective flux pdf is one which obeys (2.65) in homogeneous soil. If we insert (2.65) into (3.37), we obtain (see Problem 3.2)

$$f^r(z,t) = \frac{t}{z}\, f^f(z,t) = \frac{t\ell}{z^2}\, f^f\Big(\ell, \frac{t\ell}{z}\Big) \,, \tag{3.44}$$

where ℓ is the reference depth where the model parameters are defined. Thus, the resident pdf of the CLT lognormal model defined by (2.72) is

$$f^r(z,t) = \frac{1}{\sqrt{2\pi}\,\sigma_\ell z} \exp\Big(-\frac{\left(\ln(t\ell/z) - \mu_\ell\right)^2}{2\sigma_\ell^2}\Big) \,. \tag{3.45}$$

Figure 3.5 shows a plot of the resident pdfs for the CLT (3.45) and CDE ((3.12) divided by J_w), when the flux pdfs have been fitted to each other at $z = 50$ cm.

Note that the CDE model has a nonzero surface resident concentration, which is required to make the flux through the surface zero for times greater than zero. In contrast, the CLT model is purely convective, and solute cannot move upward, or remain at the surface unless some of the travel times in the distribution are zero. Even though these model concentrations were virtually alike when they moved past 50 cm, the CLT model solute concentration has penetrated far deeper into the soil at the time of observation in Figure 3.5.

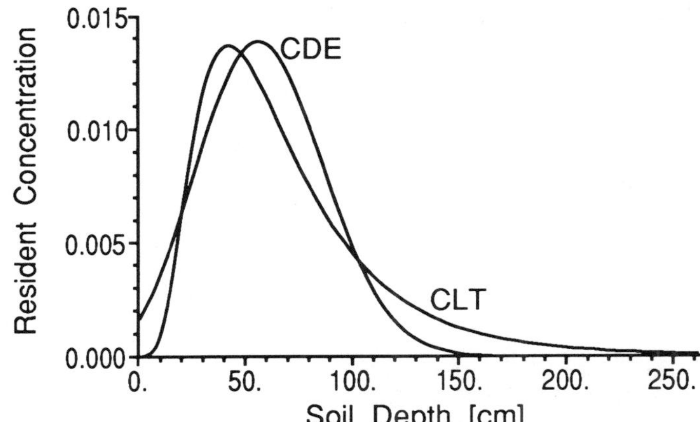

Figure 3.5: Resident pdfs of the CDE and CLT models which have been fitted as flux concentrations at $z = 50$ cm.

Example 3.5 Depth Moments of the Stochastic-Convective Model

The depth moments Z_N (3.40) of any model resident pdf which is stochastic-convective (see (3.44)) may be easily calculated in terms of the travel time moments $\mathrm{E}_\ell(t^N)$ of the flux pdf $f^f(\ell, t)$ at the reference depth ℓ as follows.
Inserting (3.44) into (3.40), we obtain

$$\begin{aligned} Z_N(t) &= \int_0^\infty z^N f^r(z,t)\,dz = \int_0^\infty z^N \frac{t\ell}{z^2} f^f\left(\ell, \frac{t\ell}{z}\right) dz \\ &= (t\ell)^N \int_0^\infty (t')^{-N} f^f(\ell, t')\,dt' = (t\ell)^N\ \mathrm{E}_\ell(t^{-N})\ , \end{aligned} \tag{3.46}$$

where the substitution $t' = t\ell/z$ has been made in the second line of (3.46).

If we again define the effective mean resident velocity $V^r_{\mathit{eff}} := dZ_1/dt$ of this process as the time derivative of the mean displacement, we calculate

$$V^r_{\mathit{eff}} = \frac{dZ_1(t)}{dt} = \ell\,\mathrm{E}_\ell\left(\frac{1}{t}\right)\ . \tag{3.47}$$

In contrast, the effective mean flux concentration velocity V^f_{eff} of a stochastic-convective process in terms of the observed first travel time moment (2.67) is

$$V^f_{\mathit{eff}} = \left(\frac{d\,\mathrm{E}_\ell(t)}{dz}\right)^{-1} = \frac{\ell}{\mathrm{E}_\ell(t)}\ . \tag{3.48}$$

Thus, even though each of these effective velocities[4] are constant, they are not the same (except for the piston flow model).

---

[4]The velocities in (3.47) and (3.48) are sometimes called Lagrangian and Eulerian velocities, respectively (Simmons, 1986b).

3.6 The Initial Value Problem

3.6.1 The Initial Value Problem in Infinite Soil

Thus far the transfer function analysis has been restricted to cases where solute is added to the inlet end of a transport volume which is initially devoid of solute. In certain applications, such as a waste spill, solute is present in the soil at $t = 0$. For such cases, we would like to find a way of representing the resident concentration for times $t > 0$ in terms of the initial profile. The problem is quite easy to solve in infinite soil (Carslaw and Jaeger, 1959), and can be represented in a mathematical form that is very similar to (2.4) representing the solution to the boundary value problem. If the initial resident concentration in the soil is

$$C_t^r(z,0) = C_0(z) \ , \tag{3.49}$$

then the resident concentration at a later time t is

$$C_t^r(z,t) = \int_{-\infty}^{\infty} C_0(z-z')\, f^r(z',t)\, dz' \ , \tag{3.50}$$

where the resident pdf $f^r(z,t)$ of the initial value problem is equal to the solute resident concentration at time t, when the initial solute concentration is a delta function $C_0(z) = \delta(z)$. The resident and flux concentrations are related as before by conservation laws ((3.2) and (3.6)).

Example 3.6 THE RESIDENT PDF OF THE CDE IN INFINITE SOIL

The solution to the resident CDE (2.36) in infinite soil for the initial condition,

$$C_l^r(z,0) = \frac{\delta(z)}{\theta} \ , \tag{3.51}$$

is obtained as follows.

We define the Fourier transform $\widetilde{C}_l^r(\lambda,t)$ of $C_l^r(z,t)$ by the integral transformation (see also Appendix A)

$$\widetilde{C}_l^r(\lambda,t) := \int_{-\infty}^{\infty} C_l^r(z,t)\, \exp(-i\lambda z)\, dz \ , \tag{3.52}$$

where $i := \sqrt{-1}$. Equation (3.52) has the inverse transform

$$C_l^r(z,t) := \frac{1}{2\pi}\int_{-\infty}^{\infty} \widetilde{C}_l^r(\lambda,t)\, \exp(i\lambda z)\, d\lambda \ . \tag{3.53}$$

If we apply the Fourier transform operation to the CDE (2.36) and the initial condition (3.51), assuming that both the flux and the concentration vanish at $\pm\infty$, we obtain (see Appendix A)

$$\frac{d\widetilde{C}_l^r(\lambda,t)}{dt} + (\lambda^2 D + i\lambda V)\, \widetilde{C}_l^r(\lambda,t) = 0 \ , \tag{3.54}$$

$$\widetilde{C}_l^r(0) = \frac{1}{\theta} \, . \tag{3.55}$$

The solution to this equation is obtained using the trial solution $\exp(\gamma t)$, which requires that $\gamma = -(i\lambda V + \lambda^2 D)$. Therefore

$$\widetilde{C}_l^r(\lambda, t) = \frac{1}{\theta} \exp(-i\lambda V t - \lambda^2 D t) \, . \tag{3.56}$$

Equation (3.56) may now be inverted with (3.53), with the result (see Example A.8)

$$C_t^r(z,t) = f^r(z,t) = \frac{1}{2\sqrt{\pi D t}} \exp\left(-\frac{(z - Vt)^2}{4Dt}\right) \, . \tag{3.57}$$

3.6.2 The Initial Value Problem of the Semi-Infinite CDE Model

The initial value problem in semi-infinite soil is more complicated than in infinite soil, because solute must now be prevented from migrating upward above the surface. In this case, (3.57) is only approximately correct for long times after the solute has migrated below the surface, for a process that obeys the CDE. The resident CDE problem is solved in Problem 3.7 for a general initial condition. In Problem 3.8 it is shown that the resident pdf of the initial value problem $C_0(z) = \delta(z)$ is the same as the resident pdf ((3.12) divided by J_w) of the boundary value problem.

3.6.3 The Initial Value Problem of the Stochastic-Convective Model

The stochastic-convective model, whose flux pdf obeys (2.65), may be solved easily for an initial resident concentration in semi-infinite soil. Because solute is not allowed to diffuse backwards toward the surface in this model, the resident pdfs for infinite and semi-infinite initial value problems are the same. Thus, if in the semi-infinite problem there is no flux through the surface, and the initial resident concentration $C_t^r(z,0) = g(z)$, then the resident concentration for $t > 0$ is given by

$$C_t^r(z,t) = \int_0^z g(z - z') \, f^r(z', t) \, dz' \, , \tag{3.58}$$

where $f^r(z,t)$[5] is the resident pdf of the initial value problem, which is defined as the resident concentration distribution at time t resulting from the addition of a delta function of solute resident concentration to the position $z = 0$ at $t = 0$.

[5] In this notation z is a random variable and t is a parameter.

In analogy to the development of the stochastic-convective flux pdf of the boundary value problem, we will define a stochastic-convective resident pdf of the initial value problem as one which obeys the following hypothesis:

> **Hypothesis 3** If a solute tracer is present in the resident fluid at $z = 0$ and $t = 0$, the probability that it will be found at a depth less than z at time t is the same as the probability that it will be found at a depth less than $z\tau/t$ at time τ.

Using the definition of a cdf (2.22), we may write hypothesis 3 as

$$P^r(z,t) = \int_0^z f^r(z',t)\,dz' = P^r\left(\frac{z\tau}{t},\tau\right) . \tag{3.59}$$

The resident pdf corresponding to the cdf in (3.59) is therefore

$$f^r(z,t) = \frac{dP^r(z,t)}{dz} = \frac{\tau}{t}\, f^r\left(\frac{z\tau}{t},\tau\right) . \tag{3.60}$$

The time τ in (3.59) and (3.60) is a reference time at which the resident pdf is measured. It is analogous to the reference depth $z = \ell$ at which the flux pdf is measured (see (2.65)). The flux concentration associated with the initial value problem is calculated from (3.60) in the usual way with (3.6). This flux concentration would correspond to an outflow concentration measurement made at a fixed depth in the soil, when the transport volume initially contained solute within it.

Problems

Problem 3.1 Calculate the resident concentration solution of the CDE corresponding to a narrow pulse input of solute directly from the differential equation and boundary conditions.

Problem 3.2 Derive (3.44) relating the resident pdf of the stochastic-convective process to the reference flux pdf.

---

[6]Problems marked with a † are more difficult.

Problem 3.3 Derive the solution for the mean and variance of the resident pdf of the CDE by inverting the formula (3.43) for the depth moments.

Problem 3.4 Derive the analytic solution of the initial value problem for the stochastic convective model (3.58) for the case of a lognormal CLT (2.72) and the initial profile $g(z) = C_0(H(z) - H(z - L))$.

Problem 3.5 Calculate the flux and resident pdfs of a porous medium made up of a fraction θ_1 through which solute moves by piston flow at a velocity $V_1 = J_w/\theta_1$, and a second fraction $\theta_2 = \theta - \theta_1$ through which it moves at V_2, where J_w is the steady water flux rate. Assume that there is no interaction between the two zones. Calculate the flux and resident mean velocities of the system.

†**Problem 3.6** The two region flow system of the previous problem may be extended to the case where there is rate limited mass transfer between the two zones, such that the flux concentrations in each zone obey the equations

$$\theta_1 \frac{\partial C_1^f}{\partial t} + J_w \frac{\partial C_1^f}{\partial z} = \alpha(C_2^f - C_1^f) \,, \tag{3.61}$$

$$\theta_2 \frac{\partial C_2^f}{\partial t} + J_w \frac{\partial C_2^f}{\partial z} = \alpha(C_1^f - C_2^f) \,, \tag{3.62}$$

and the boundary and initial conditions

$$C_i^f(z, 0) = 0 \quad ; \quad i = 1, 2 \,, \tag{3.63}$$

$$C_i^f(0, t) = \frac{\theta_i}{\theta}\delta(t) \quad ; \quad i = 1, 2 \,. \tag{3.64}$$

Calculate the mean and variance of the travel time of the total flux concentration $C^f = C_1^f + C_2^f$.

†**Problem 3.7** Calculate the general solution to the resident CDE for zero flux input of solute, steady state water flow, and the initial condition

$$C_l^r(z, 0) = g(z) \,. \tag{3.65}$$

Problem 3.8 Evaluate the result in Problem 3.7 for the special case $g(z) = \delta(z)/\theta$.

Problem 3.9 Evaluate the result in Problem 3.7 for the special case $g(z) = C_0((H(z) - H(z - L))$.

Chapter 4

Stochastic Stream Tube Modeling

In stochastic stream tube modeling, the soil chemical transport process at the field scale is modeled by using a local process theory, like a differential equation solution, to represent how a chemical moves in a given stream tube. Transport through the stream tube is represented by a simple process model whose parameters are constant, and each tube is isolated from adjacent tubes in the field (i.e. no solute exchange). Solute concentration in the stream tube is usually expressed one-dimensionally as a function of z and t, so that the stream tubes are like parallel soil columns.

Area-averaged field-scale solute concentrations are generated in this model by assuming that the parameters of the local stream tube model are random variables described by probability distributions, but having no spatial correlation structure. The field-averaged concentration is therefore calculated as an expectation or ensemble average over the probability distributions.

The stochastic stream tube model can be expressed as a transfer function quite easily, allowing measured outflow flux concentrations to be interpreted in terms of the spatial variability of the parameters of the local model used to represent transport through the stream tubes. Formally, the approach can be summarized as follows:

Assume that the soil consists of a series of noninteracting stream tubes, within which solute moves as a function of z and t according to some simple process model. The parameters $\lambda_1, \dots, \lambda_N$ of the process model are constants *within* a given tube, and uncorrelated, random variables *between* tubes. Their distribution is described by the joint probability density function $f(\lambda_1, \dots, \lambda_N)$. All stream tubes connect to the surface, and are subjected to the same upper boundary condition.

Denote the solution (flux or resident solute concentration) of the process model inside a particular stream tube by $C(z,t;\lambda_1,\ldots,\lambda_N)$. The concentration in the ensemble of stream tubes may then be regarded as a random process $\mathbf{C}(z,t)$. The mean concentration is given by the ensemble average

$$\begin{aligned}\overline{C}(z,t) &:= \mathrm{E}(\mathbf{C}) \\ &= \int_{-\infty}^{\infty}\cdots\int_{-\infty}^{\infty} C(z,t;\lambda_1,\ldots,\lambda_N)\, f(\lambda_1,\ldots,\lambda_N)\, d\lambda_1\cdots\lambda_N \ ,\end{aligned} \tag{4.1}$$

where the integrals range over all values of the random parameters $\lambda_1,\ldots,\lambda_N$.

Example 4.1 PISTON FLOW THROUGH PARALLEL SOIL COLUMNS

The simplest chemical process law we have used to describe the transport of a mobile solute is the so-called piston flow model (see Example 2.1), where the solute is carried by mass flow without undergoing diffusive or dispersive mixing. If a particular soil column has water flowing through it at a constant velocity V, then the piston flow model solution for the flux concentration at depth z and time t when a narrow pulse (delta function) of solute has been added to the surface at time zero is

$$C^f(z,t;V) = \delta(t - z/V) \ , \tag{4.2}$$

where V is a parameter which is constant in a particular soil column but is distributed randomly over the field. Thus, V is described by a pdf $f_V(V)$, where $V > 0$. The field-averaged flux concentration is, according to (4.1),

$$\begin{aligned}\overline{C}^f(z,t) = \mathrm{E}(\mathbf{C}^f(z,t)) &= \int_0^{\infty} \delta(t - z/V)\, f_V(V)\, dV \\ &= \int_0^{\infty} \delta(t-y)\, f_V(\frac{z}{y})\,\frac{z\,dy}{y^2} = \frac{z}{t^2}\, f_V(\frac{z}{t}) \ ,\end{aligned} \tag{4.3}$$

where the substitution $V = z/y$ has been used in the integral.

Since $\overline{C}^f(z,t)$ is the area-averaged solute flux concentration at the outflow depth z in response to a delta function input at the surface, it is by definition equal to the travel time flux pdf for this model. Thus,

$$f^f(z,t) = \frac{z}{t^2} f_V(\frac{z}{t}) \ . \tag{4.4}$$

This example is worth some discussion, because it illustrates how the transfer function approach can be used to test the assumptions of process models. The example above used a physical model (the parallel soil columns) to predict an observable event (the field average concentration resulting from a narrow pulse input of solute). If this simple process model is correct, then solute moves through the field in the vertical direction only, at a local velocity V. In order to validate the model, one must measure the velocity distribution independently,[1] or predict it from measurements of the distribution of soil properties which influence V. The next example provides an illustration of the latter approach.

Example 4.2 THE RANDOM WATER CONTENT MODEL

In the previous example, each soil column was assumed to be transporting solute at a different, but constant, velocity V when the soil was at a steady water flux J_w. If we assume that all of the wetted pore space is part of the solute transport volume, then we can write

$$V = \frac{J_w}{\theta} , \tag{4.5}$$

where θ is the volumetric water content, assumed to be constant within a column and a random variable over the field. With this relation, $f_V(V)$ in (4.4) can be predicted from the observed variation of θ. Moreover, each column has a single travel time

$$t = \frac{z\theta}{i_0} \tag{4.6}$$

so that the attributes of the travel time pdf $f^f(z,t)$ can be calculated directly from the distribution of θ. If we assume that all of the distributions are lognormal, the calculations become quite simple if the following theorem is used (Aitcheson and Brown, 1961).

> **Theorem** If X is a lognormally distributed random variable with distribution parameters $\mathrm{E}(\ln(X)) = \mu_X$ and $\mathrm{Var}(\ln(X)) = \sigma_X^2$, then $Y := aX^b$ is lognormally distributed with parameters $\mathrm{E}(\ln(Y)) = \mu_Y = \ln(a) + b\mu_X$ and $\mathrm{Var}(\ln(Y)) = \sigma_Y^2 = b^2\sigma_X^2$. (The proof of this theorem is given in Problem 4.4.)

Application of this theorem to (4.6) gives us the result that if the travel time is lognormally distributed, then θ should be lognormally distributed and have the same σ^2 as t. Since the coefficient of variation CV of a lognormal distribution is equal to (Aitcheson and Brown, 1961)

$$\mathrm{CV} = \sqrt{\exp(\sigma^2) - 1} \tag{4.7}$$

then the observed CV of θ should be comparable to the CV of the travel time distribution $f^f(z,t)$.

In the field studies conducted to date where both quantities were measured, the travel time CV has been of the order of 0.50 or higher, whereas the measured CV of θ is usually 0.2 or less (Jury, 1985). Thus, not surprisingly, much of the travel time variability cannot be explained by simple one dimensional piston flow, and the solute transport volume is not equal to the total volume occupied by the water.

---

[1] One could not "measure" the velocity distribution by fitting $f_V(V)$ to the outflow pdf $f^f(z,t)$. This would merely parameterize the flux pdf. The process model assumption which must be tested is the hypothesis that the velocities are one dimensional in the z direction. The validity of this assumption could not be established only by looking at outflow curves.

4.1 Solute Transport and Adsorption

In all of the discussion thus far, it has been assumed that the solutes moving through the transport volume do not interact with the soil solid phase. This assumption might be expected to apply reasonably well to inorganic anions such as Cl^-, NO_3^-, or Br^-, or to certain soluble organic compounds moving through soil with low organic matter. However, many chemicals of agricultural or environmental interest are attracted to soil mineral or organic surfaces, and are slowed down to varying degrees compared to the movement of a mobile tracer. In this section, we will examine the adsorption reaction as represented by different process models, and its effect on solute transport. In all cases, we will assume that the total resident concentration is partitioned into two terms

$$C_t^r = \rho_b C_a^r + \theta C_l^r \ , \tag{4.8}$$

where ρ_b is the soil dry bulk density and C_a^r is the adsorbed concentration expressed in units of mass of solute adsorbed per mass of dry soil.

4.2 Linear Equilibrium Adsorption

The simplest adsorption model is the linear equilibrium adsorption relationship

$$C_a^r = K_d C_l^r \ , \tag{4.9}$$

where K_d is the distribution coefficient (Hamaker and Thompson, 1972), assumed to be constant for a given set of soil conditions (temperature, pH, organic C, etc.). In this model, any solute added to the dissolved phase instantaneously partitions into the adsorbed phase according to (4.9). Thus, the ratio between the total resident concentration C_t^r and the mass per soil volume θC_l^r residing in the mobile phase may be calculated with (4.8) and (4.9),

$$\frac{C_t^r}{\theta C_l^r} = 1 + \frac{\rho_b K_d}{\theta} =: R \ , \tag{4.10}$$

where R is called the retardation factor. Thus, if an adsorbing tracer is added to the dissolved phase, it has only a probability $1/R$ of remaining there at any given time. Therefore, an adsorbing chemical will take R times as long as a nonadsorbing chemical to travel a given distance, as long as the two compounds move through the same transport volume in the dissolved phase.

Since the mobile tracers used to measure or model the flux pdf are not adsorbed, (4.10) provides a means of estimating the travel time pdf of a linearly adsorbing tracer in terms of the pdf of a mobile tracer. According to (4.10), an adsorbing chemical will spend $1/R$ as much time in the dissolved phase as a mobile one. If we assume that the mobile and adsorbing tracers move through the same water flow pathways in the soil, then the probability $P_a^f(z,t)$ that an adsorbing tracer added to the surface at $t = 0$ will reach depth z in a time less than or equal to t is given by

$$P_a^f(z,t) = P_m^f(z,t/R) , \tag{4.11}$$

where $P_m^f(z,t/R)$ is the travel time flux cdf of the mobile tracer, and the subscripts a and m refer to the adsorbing and mobile tracers. Then, using the definition (2.24) relating the pdf to the cdf, we obtain

$$f_a^f(z,t) = \frac{1}{R} f_m^f\left(z, \frac{t}{R}\right) . \tag{4.12}$$

Equation (4.12) is valid for any process representation of the mobile flux pdf, provided that the mobile and adsorbing tracers move through the same regions of the transport volume with the same probability when in the dissolved phase.[2]

Example 4.3 LINEAR ADSORPTION AND THE CDE

The CDE model (2.37) can be shown to obey the flux pdf relation (4.12) when the chemical adsorbs to equilibrium as follows.

If we combine the solute conservation equation (3.2) with the CDE solute flux (2.34), and the total resident concentration (4.8) subject to linear, equilibrium adsorption (4.10) we obtain the generalized form of the CDE model

$$R\frac{\partial C_l^r}{\partial t} + V\frac{\partial C_l^r}{\partial z} - D\frac{\partial^2 C_l^r}{\partial z^2} = 0 , \tag{4.13}$$

where $R = 1$ for the mobile tracer. Since the CDE relation (3.14) between flux and resident concentrations is still valid for an adsorbing tracer (because (2.34) still describes the solute flux), the flux concentration also satisfies the generalized CDE

$$R\frac{\partial C^f}{\partial t} + V\frac{\partial C^f}{\partial z} - D\frac{\partial^2 C^f}{\partial z^2} = 0 . \tag{4.14}$$

If we let $t' = t/R$ in (4.14), we obtain the CDE for a mobile tracer (2.37) in the new time frame t'. Therefore, if

$$C^f(0,t') = \delta(t') \tag{4.15}$$

then

$$C^f(z,t') = f_m^f(z,t') \tag{4.16}$$

[2]In other words, the adsorbing chemical moves through exactly the same volume as the mobile chemical, but gets slowed down by the adsorption reaction.

is the travel time pdf of the mobile solute. Finally, since $t' = t/R$

$$f_a^f(z,t) = \frac{1}{R} f_m^f\left(z, \frac{t}{R}\right) , \tag{4.17}$$

where the extra $1/R$ is required for normalization so that $f_m^f(z,t)$ will have unit area.

4.3 Rate Limited Adsorption

The dissolved and adsorbed phases are physically separated. When the phases are displaced from equilibrium at some point in the soil, say by the sudden addition of adsorbing solute to the inlet end of a soil column, then restoration of equilibrium is not instantaneous, but occurs over a characteristic time period. This time, which can be regarded as the characteristic time scale of the adsorption process, is determined by the speed with which solute molecules diffuse through the liquid in the vicinity of the adsorbing interfaces. If the adsorption time scale is much shorter than the time of contact between the incoming solution and the adsorbing surfaces, then equilibrium will eventually be restored and a partitioning law such as (4.9) can be used to relate the concentrations in the separate phases. However, if the solution is moving so rapidly that it leaves the vicinity of an adsorbing surface before equilibrium is restored, then the partitioning law will be inaccurate, and a different process representation (rate limited adsorption) is needed to relate the phases.

When the solute is undergoing rate limited adsorption, the relationship between the mobile and sorbed phases is time-dependent. A common way to represent this relationship is to assume that the mass transfer rate between the two phases is proportional to the deviation from equilibrium (Lindstrom and Boersma, 1971; Van Genuchten et al., 1974)

$$\rho \frac{\partial C_a^r}{\partial t} = \alpha (K_d C_l^r - C_a^r) . \tag{4.18}$$

Equations (4.8) and (4.18) constitute a model for the resident concentration in the soil. The next example shows how this can be used with the CDE equation.

Example 4.4 RATE LIMITED ADSORPTION AND THE CDE

The CDE flux equation, expressed in terms of resident fluid concentration

$$J_s = J_w C^f = -\theta D \frac{\partial C_l^r}{\partial z} + J_w C_l^r \tag{4.19}$$

can be combined with the solute conservation relation (3.2) and the resident concentration (4.8) to produce

$$\theta \frac{\partial C_l^r}{\partial t} + \rho_b \frac{\partial C_a^r}{\partial t} - \theta D \frac{\partial^2 C_l^r}{\partial z^2} + J_w \frac{\partial C_l^r}{\partial z} = 0 , \tag{4.20}$$

where we assume steady water flux and uniform soil properties. This equation has two unknowns, the two resident phase concentrations. The rate limited phase concentration relation (4.18) provides the second equation needed to solve for the two unknowns. If there is no solute present in the soil initially, the Laplace transform of (4.18) may be solved for $\widehat{C}_a^r$

$$\widehat{C}_a^r = \frac{\alpha K_d \widehat{C}_l^r}{\alpha + \rho_b s} . \tag{4.21}$$

The corresponding Laplace transform of (4.20), divided by θ, is

$$D\frac{d^2\widehat{C}_l^r}{dz^2} - V\frac{d\widehat{C}_l^r}{dz} - s\widehat{C}_l^r - s\frac{\rho_b}{\theta}\widehat{C}_a^r = 0 . \tag{4.22}$$

After insertion of (4.21) into (4.22), we obtain

$$D\frac{d^2\widehat{C}_l^r}{dz^2} - V\frac{d\widehat{C}_l^r}{dz} - g(s)\widehat{C}_l^r = 0 , \tag{4.23}$$

where the function $g(s)$ is equal to

$$g(s) = s + \frac{s\alpha(R-1)}{\alpha + \rho_b s} \tag{4.24}$$

and R is given by (4.10).

For an upper boundary condition consisting of a narrow-pulse input of solute flux

$$- D\frac{\partial C_l^r}{\partial z} + VC_l^r = VC^f = V\delta(t) \qquad \text{at} \quad z = 0 , \tag{4.25}$$

whose Laplace transform is

$$- D\frac{d\widehat{C}_l^r}{dz} + V\widehat{C}_l^r = V\widehat{C}^f = V \qquad \text{at} \quad z = 0 , \tag{4.26}$$

the solution to (4.23) is given by

$$\widehat{C}_l^r = \frac{2}{1+\zeta} \exp\Big(\frac{Vz}{2D}(1-\zeta)\Big) , \tag{4.27}$$

where

$$\zeta := \sqrt{1 + 4g(s)D/V^2} . \tag{4.28}$$

Therefore, the Laplace transform of the total resident concentration is—using (4.8), (4.21) and (4.27)

$$\widehat{C}_t^r = \frac{2\theta g(s)}{(1+\zeta)s} \exp\Big(\frac{Vz}{2D}(1-\zeta)\Big) . \tag{4.29}$$

Finally, the flux concentration is given by

$$\widehat{C}^f = \widehat{C}_l^r - \frac{D}{V}\frac{d\widehat{C}_l^r}{dz} = \exp\Big(\frac{Vz}{2D}(1-\zeta)\Big) , \tag{4.30}$$

Notice that (4.30) could have been obtained directly, since the flux concentration $\widehat{C}^f$ also obeys (4.23).

Since (4.30) is the travel time pdf for this process, we may calculate the travel time moments for this model by (2.29). As shown in Problem 4.9, we obtain

$$\mathrm{E}_z(t) = Rz/V \tag{4.31}$$

$$\mathrm{Var}_z(t) = \frac{2R^2Dz}{V^3} + \frac{2\rho_b z(R-1)}{\alpha V} . \tag{4.32}$$

Thus, much of the variance or solute mixing in this model is caused by the rate limited adsorption process. Figures 4.1 and 4.2 show solutions for the flux and resident concentrations for a compound undergoing rate limited adsorption with an equilibrium $R = 3$, for three values of the dimensionless rate parameter $W = \alpha\ell/V$. For extremely small values of the rate parameter, the compound behaves like it is not adsorbing at all and for large values it behaves like a compound adsorbing to equilibrium. The intermediate value of the rate parameter produces an outflow curve with unique features like a long outflow tail and a poorly leached zone near the surface.

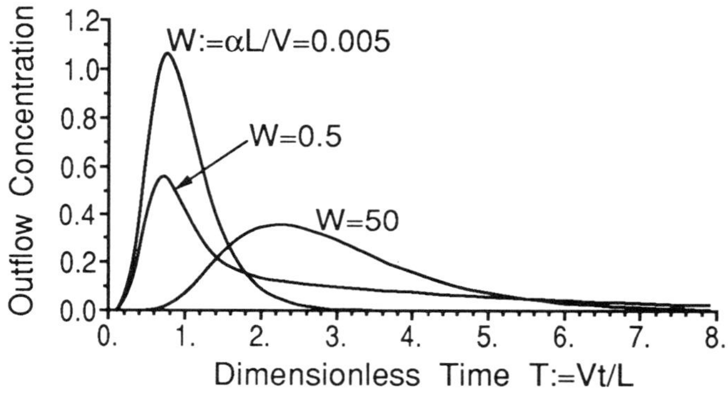

Figure 4.1: *Flux concentration pdfs for a solute undergoing rate limited adsorption.*

4.4 Spatial Variability and Equilibrium Adsorption

If a tracer adsorbs linearly to equilibrium, but the retardation factor R changes at different locations on the field, then (4.17) is no longer valid. This problem may be solved by the stochastic stream tube method (4.1). According to (4.10),

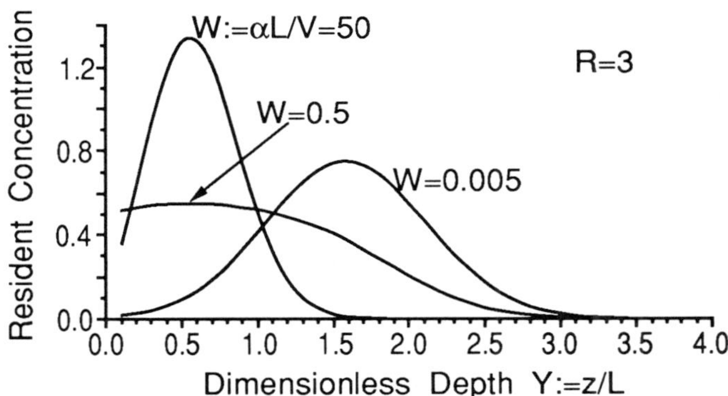

Figure 4.2: *Total resident concentrations for a solute undergoing rate limited adsorption.*

the travel time t_a of an adsorbing solute which has a retardation factor is related to the travel time t_m of a mobile solute which does not adsorb by

$$t_a = Rt_m \ , \tag{4.33}$$

where we may regard R and t_m as random variables which are constant in a given stream tube. Thus, by (4.1) the flux pdf of the field-averaged concentration is

$$f^f(z,t) = \int_0^\infty \int_0^\infty \delta(t - Rt_m)\, f(R,t_m)\, dR\, dt_m \ , \tag{4.34}$$

where $f(R,t_m)$ is the joint pdf of R and t_m. Equation (4.34) may be evaluated analytically if $f(R,t_m)$ is a bivariate lognormal distribution

$$f(R,t_m) = \frac{1}{2\pi\,\sigma_R\,\sigma_m\, Rt_m\sqrt{1-\rho^2}} \ \exp\Bigl(-\frac{y_R^2 - 2\rho\, y_R\, y_m + y_m^2}{2(1-\rho^2)}\Bigr) \ , \tag{4.35}$$

where

$$y_m = \frac{\ln(t_m) - \mu_m}{\sigma_m} \tag{4.36}$$

$$y_R = \frac{\ln(R) - \mu_R}{\sigma_R} \tag{4.37}$$

are zero mean, unit variance, random normal variates, and

$$\rho := \mathrm{E}(y_R\, y_m) = \int_0^\infty \int_0^\infty y_m\, y_R\, f(R,t_m)\, dR\, dt_m \tag{4.38}$$

is the correlation coefficient. In this case, (4.34) reduces to (see Problem 4.2)

$$f^f(z,t) = \frac{1}{\sqrt{2\pi}\sigma t} \exp\left(-\frac{(\ln(t) - \mu)^2}{2\sigma^2}\right) , \tag{4.39}$$

where

$$\mu := \mu_R + \mu_m \tag{4.40}$$

$$\sigma := \sqrt{\sigma_R^2 + 2\rho\,\sigma_R\,\sigma_m + \sigma_m^2} \ . \tag{4.41}$$

Thus, the adsorbing solute also has a lognormal travel time pdf. This model is explored in the next example.

Example 4.5 SPATIAL VARIABILITY OF PESTICIDE ADSORPTION

A field experiment is performed in which a mobile tracer (chloride) is added as a pulse to the surface, and the mobile travel time pdf is measured by monitoring the movement of the pulse past a fixed depth $z = \ell$ with solution samplers. At the same time, soil samples are taken at a number of locations and are used to determine the distribution of equilibrium adsorption retardation factors R for a pesticide (bromacil). Both R and t_m are found to be lognormal. Three models are used to predict the travel time pdf of the bromacil between the soil surface and depth ℓ relative to the travel time pdf of the chloride:

- Model A: The bromacil adsorbs to equilibrium with a single R equal to the mean R_0 of the distribution.
- Model B: The bromacil adsorbs to equilibrium with a random R which is different in each stream tube but is uncorrelated with the travel time of the chloride in that tube.
- Model C: The bromacil adsorbs to equilibrium with a random R which is different in each stream tube and is perfectly correlated with the travel time of the chloride in that tube.

It follows from (4.40)–(4.41) and the properties (2.77) and (4.7) of the lognormal distribution that the mean and CV of the bromacil travel time distribution are predicted to be the following according to the three models:

Model A

$$\mathrm{E}(t_a) = R_0\ \mathrm{E}(t_m) \tag{4.42}$$

$$\mathrm{CV}(t_a) = \mathrm{CV}(t_m) = \sqrt{\exp(\sigma_m^2) - 1} \tag{4.43}$$

Model B

$$\mathrm{E}(t_a) = R_0\ \mathrm{E}(t_m) \tag{4.44}$$

$$\mathrm{CV}(t_a) = \sqrt{\exp(\sigma_R^2 + \sigma_m^2) - 1} \tag{4.45}$$

Model C

$$\mathrm{E}(t_a) = R_0\ \mathrm{E}(t_m)(1 + \mathrm{CV}(R)\ \mathrm{CV}(t_m)) \tag{4.46}$$

$$\mathrm{CV}(t_a) = \sqrt{\exp((\sigma_R + \sigma_m)^2) - 1} \tag{4.47}$$

Figure 4.3 shows the outflow concentrations predicted for models A–C for a solute with a spatially variable retardation factor R. The difference in the predicted outflow shapes is quite significant, even between Models B and C which use the same data. This difference illustrates the importance of correlations between adsorption and mobile travel time; unfortunately, correlation coefficients are difficult to measure in the field.

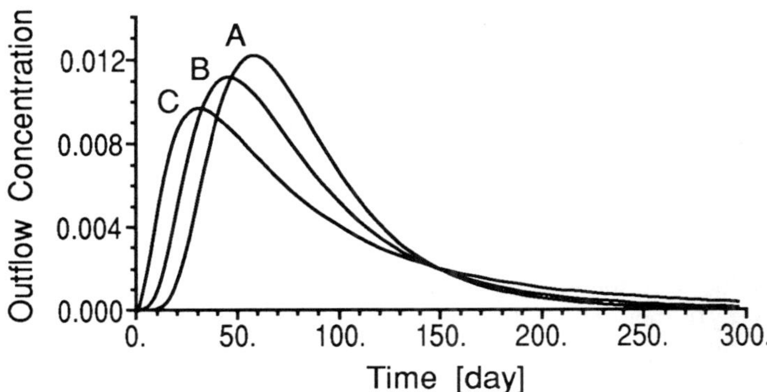

Figure 4.3: *Predicted outflow concentrations of models A–C for a solute with a retardation factor R which has a coefficient of variation of 0.40, moving through a soil with a mobile travel time pdf which has a CV of 0.55.*

El Abd (1984), and Jury et al. (1986) measured retardation factors of napropamide in soil taken from 36 locations over a 0.64 ha field. They found that the distribution of R was skewed, with a CV of 0.31. No apparent correlation was found between R and t_m in individual cores, but the data suggested that part of the pesticide was moving without adsorbing to equilibrium.

It is also possible to construct simple solutions for the solute travel time pdf when the adsorption coefficient varies in the vertical direction. This topic is addressed in Chapter 5.

4.5 Moments of Stream Tube Models

If (4.1) is applied to the narrow pulse input boundary condition, then both the local flux concentration $C^f(z,t;\lambda_1,\ldots,\lambda_N)$ inside the integral and the mean concentration $\overline{C}^f(z,t)$ may be regarded as travel time pdfs. Thus, in this case, we may write (4.1) as

$$\begin{aligned} \mathrm{E}(\mathbf{C}) &= f^f(z,t) \\ &= \int_{-\infty}^{\infty}\cdots\int_{-\infty}^{\infty} f^f(z,t;\lambda_1,\ldots,\lambda_N)\, f(\lambda_1,\ldots,\lambda_N)\, d\lambda_1\cdots d\lambda_N \,. \end{aligned} \tag{4.48}$$

Therefore, the Nth travel time moment of the system may be written as

$$\mathrm{E}_z(t^N) \;=\; \int_0^{\infty} t^N\, f^f(z,t)\, dt$$

$$= \int_{-\infty}^{\infty} \cdots \int_{-\infty}^{\infty} \mathrm{E}_z(t^N; \lambda_1, \ldots, \lambda_N)\, f(\lambda_1, \ldots, \lambda_N)\, d\lambda_1 \cdots d\lambda_N \; . \tag{4.49}$$

Equation (4.49) shows that the Nth travel time moment of the mean travel time pdf is equal to the ensemble average of the local Nth moment. This result can be used to simplify the calculation of the travel time moments of the mean pdf in certain problems, and also to clarify the role of various factors contributing to the values of the moments.

Example 4.6 MOMENTS OF THE PARALLEL SOIL COLUMN MODEL

The parallel soil column model described locally by (4.2) may be used to illustrate (4.49). The Nth moment of (4.2) (which is a normalized local pdf because it is the outflow flux concentration in response to a delta function input) is, from (2.54)

$$\mathrm{E}_z(t^N; V) = \left(\frac{z}{V}\right)^N . \tag{4.50}$$

Thus, the average Nth moment for the field is, by (4.49)

$$\mathrm{E}_z(t^N) = z^N \; \mathrm{E}_V\left(\left(\frac{1}{V}\right)^N\right) , \tag{4.51}$$

where $\mathrm{E}_V(.)$ is the expectation or ensemble average over $f_V(V)$. In particular, the mean and variance are

$$\mathrm{E}_z(t) = z \; \mathrm{E}_V\left(\frac{1}{V}\right) \tag{4.52}$$

$$\mathrm{Var}_z(t) = z^2 \; \mathrm{Var}_V\left(\frac{1}{V}\right) . \tag{4.53}$$

Since these moments are determined from the flux pdf, we may calculate the effective flux velocity and dispersivity of this model using the definitions (2.57)–(2.58) for the moments of the CDE.

$$V^f_{\mathit{eff}} = \left(\mathrm{E}_V\left(\frac{1}{V}\right)\right)^{-1} \tag{4.54}$$

$$\lambda^f_{\mathit{eff}} = \frac{D^f_{\mathit{eff}}}{V^f_{\mathit{eff}}} = \frac{z}{2} \; \mathrm{CV}^2_V\left(\frac{1}{V}\right) , \tag{4.55}$$

where $\mathrm{CV}_V(1/V)$ is the coefficient of variation of the inverse of the local velocity. Thus, there are two important results illustrated in this analysis. First, the mean flux velocity of the area-averaged pulse is the harmonic mean of the local velocities, and second, the field-averaged dispersivity, representing the spreading of the average pulse, increases linearly with distance.

4.6 Stream Tube Modeling and Variability Analysis

Stream tube modeling may be useful in the analysis of the components of spatial variability of a process which has randomness in it arising from several sources. This is explored in the next example.

Example 4.7 SOLUTE TRANSPORT UNDER SPATIALLY VARIABLE IRRIGATION

A field experiment is performed in which a narrow pulse of solute is added to the surface and leached at a spatially variable, but time-invariant rate by a sprinkler system. A travel time pdf is constructed at $z = \ell$ from solution sampler measurements. Both the travel time pdf and the pdf describing the irrigation rate are lognormal. How much of the variability of the travel time pdf was caused by the soil?

A simple answer is possible if the following assumptions are made: (i) the soil has a unique pdf as a function of net applied water I under steady-state irrigation (see Hypothesis 1 and (2.63)); (ii) the irrigation rate i and the net applied water required to reach $z = \ell$ are uncorrelated (which is reasonable if there is no ponding), and (iii) the irrigation rate has no spatial correlation.

With these assumptions the pdf for the travel time to depth z through a stream tube is simply

$$C^f(z,t;I,i) = f^f(z,t;I,i) = \delta\left(t - \frac{I}{i}\right) , \tag{4.56}$$

where I and i are treated as constant parameters in a stream tube and random variables over the field. Then, after inserting (4.56) into (4.1), changing the I variable to $y = I/i$ and integrating over y, we obtain

$$f^f(z,t) = \int_0^\infty i\, f_I(z, i\, t)\, f_i(i)\, di \ . \tag{4.57}$$

Equation (4.57) is valid regardless of how I and i are distributed, provided that they are independent. If they are both lognormal and independent, then $f(z,t)$ is also lognormal with parameters (see Problem 4.2)

$$\mu_t = \mu_I - \mu_i \tag{4.58}$$

$$\sigma_t^2 = \sigma_I^2 + \sigma_i^2 \ . \tag{4.59}$$

Figure 4.4 shows predicted travel time pdfs for a given $\mathrm{CV}_I = 0.5$ of the soil pdf as a function of the coefficient of variation CV_i of the irrigation rate. As seen from the figure, even a relatively large CV_i of 0.20 produces only a small change in the soil travel time pdf, which is dominated by the intrinsic variability of the soil (Jury, 1982).

4.7 Resident PDFs of Stream Tube Models

The parallel soil column model with a variable velocity distribution produced a flux pdf given by (4.4). Thus, using (3.37), one can show that the resident pdf

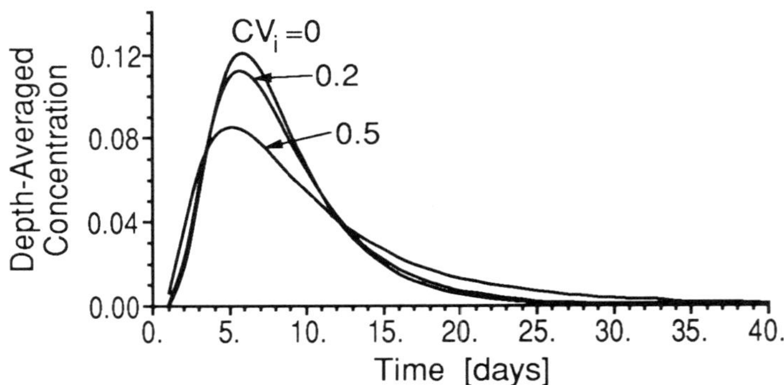

Figure 4.4: Predicted travel time pdfs to $z = 30$ cm calculated with (4.57) using the intrinsic soil pdf ($\mu_\ell = 2.0$, $\sigma_\ell = 0.5$, $\ell = 30$ cm) reported in Butters et al. (1989) for three values of the irrigation rate CV$_i$.

for this model is (see Problem 4.5)

$$f^r(z,t) = \frac{1}{t}\, f_V\left(\frac{z}{t}\right) . \tag{4.60}$$

The first depth moment of this pdf is therefore

$$Z_1 := \mathrm{E}(z) = \int_0^\infty \frac{z}{t}\, f_V\left(\frac{z}{t}\right) dz = t\int_0^\infty V f_V(V)\, dV = t\, \mathrm{E}_V(V) , \tag{4.61}$$

where we have used the substitution $V = z/t$. Therefore, the average velocity V^r_{eff} of the pulse, calculated from observations of the resident concentration, is the slope of the average position over time[3]

$$V^r_{\mathit{eff}} = \mathrm{E}_V(V) . \tag{4.62}$$

Equation (4.62) states that the mean resident concentration velocity of a pulse moving in a spatially variable velocity field is the mean of the velocity distribution. However, if we calculate the mean travel time to depth z with the flux pdf (4.4), we obtain the result given in (4.54). Therefore, the average velocity V^f_{eff} of the pulse, calculated from observations of the flux concentration, is the inverse slope of the average travel time over depth, or

$$V^f_{\mathit{eff}} = \frac{1}{\mathrm{E}_V(\frac{1}{V})} . \tag{4.63}$$

[3]The average velocity is not time dependent in this model because solute must move downward in each stream tube.

The next example will illustrate the differences between the flux and resident velocities quantitatively.

Example 4.8 SOLUTE MOVEMENT IN A LOGNORMAL VELOCITY FIELD

Assuming that the velocity distribution $f_V(V)$ (for V in cm d^{-1}) of the parallel soil column model discussed above is lognormal with $\mu = 1$ and $\sigma = 0.75$, we wish to calculate the mean flux and resident velocities of a narrow pulse added to the soil surface.

In Problem 2.4, it was shown that if x is lognormal, then

$$\mathrm{E}(x^N) = \exp\Big(N\mu + \frac{N^2\sigma^2}{2}\Big) . \tag{4.64}$$

Therefore, from (4.62)–(4.63)

$$\begin{aligned} V^r_{eff} &= \exp(\mu + \sigma^2/2) = 3.60 \\ V^f_{eff} &= \frac{1}{\exp(-\mu + \sigma^2/2)} = 2.05 . \end{aligned}$$

Thus, in this example, the mean velocity of the pulse is perceived to be almost 50% slower when viewed as a flux concentration (e.g. by solution samplers at a fixed depth) than as a resident concentration (e.g. by soil coring at a fixed time).[4]

4.8 Solute Transport with first order Decay

Thus far, we have limited our discussion of solute transport to cases where the solute does not undergo reactions in the soil. A simple reaction which can be easily incorporated into our description of solute movement through a transport volume is a first order decay process, where the rate of disappearance of the solute mass per volume is proportional to the total resident concentration. In that case, the solute conservation equation becomes

$$\frac{\partial C^r_t}{\partial t} + J_w \frac{\partial C^f}{\partial z} + \mu C^r_t = 0 , \tag{4.65}$$

where $\mu = \ln(2)/\tau_{1/2}$ is the first order decay constant, and $\tau_{1/2}$ is the half life. The Laplace transform of (4.65) for the case of zero initial solute resident concentration is

$$(s + \mu)\widehat{C}^r_t + J_w \frac{d\widehat{C}^f}{dz} = 0 , \tag{4.66}$$

[4]This result may explain why Butters et al. (1989) found the vertical mean velocity of a pulse averaged over 0.64 ha to be almost twice as slow as the piston flow velocity predicted from the mean water content profile and the steady water flux.

which is identical to the Laplace transform of the solute conservation equation (3.2) for an inert solute, except that $s \to s + \mu$. Thus, by the shifting theorem (A.33) for Laplace transforms (Appendix A), the solution for the flux concentration in response to a narrow pulse (delta function) input of solute under first order decay may be written as

$$C^f(z,t) = f^f(z,t;\mu) = \exp(-\mu\, t)\, f^f(z,t;\mu = 0)\ , \tag{4.67}$$

where $f^f(z,t;\mu = 0)$ is the travel time pdf of a nonreactive chemical. In this case, the impulse response function $f^f(z,t;\mu)$ expresses the distribution of the travel times for those solutes that reach the outflow end without undergoing first order decay during transport. Therefore, it is not normalized, because not all of the unit applied mass reaches the outflow end. Equation (4.67) is valid for any process model.

Figure 4.5 shows a plot of the travel time pdf (4.67) for various values of the decay constant μ, using a Fickian pdf (2.59) to represent the travel time of the nondecaying chemical.

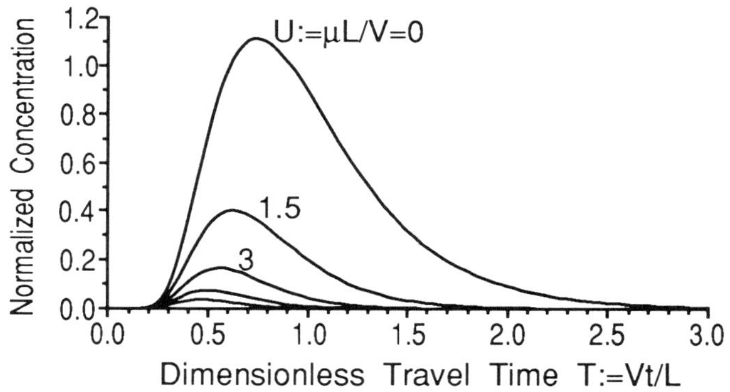

Figure 4.5: *Travel time pdf of a chemical undergoing first order decay, for a Fickian process with $P := \ell V/D = 10$, for various values of the dimensionless decay constant $U = \mu\ell/V = 0,\ 1.5,\ 3.0,\ 4.5,\ 6.0$.*

4.9 Simultaneous Transport, Adsorption, and first order Decay

A chemical which is undergoing simultaneous linear adsorption and first order decay during transport under steady state water flux at a rate J_w has a travel time pdf which may be written as (using (4.12) and (4.67))

$$f^f(z,t) = \frac{\exp(-\mu\, t)}{R}\, f^f_m\Big(z, \frac{t}{R}\Big)\;. \tag{4.68}$$

Since many soils have relatively shallow biologically active zones, a pesticide or other hazardous organic chemical that migrates below the surface soil layer where degradation processes are occurring can represent a potential threat to underlying ground water.

4.9.1 Estimation of the Mean Residual Mass Fraction (RMF)

We can estimate the residual mass fraction RMF (Jury and Gruber, 1989) of the solute mass added to the surface which survives the passage through the zone of degradation, since it is merely equal to the cumulative probability $P^f(z,\infty)$ that a solute molecule reaches the outflow end of the transport volume—the biologically active zone—in a finite travel time. Thus, using (4.68), the RMF may be written as

$$\mathrm{RMF} = P^f(z,\infty) = \int_0^\infty f^f(z,t)\, dt = \frac{1}{R}\int_0^\infty \exp(-\mu t)\, f^f_m\Big(z, \frac{t}{R}\Big)\, dt\;. \tag{4.69}$$

We may incorporate the influence of the water flux rate into (4.69) by transforming from time to cumulative net applied water or drainage $I = J_w t$, and by assuming, as we did in Chapter 2 (2.63), that the soil travel time pdf for a mobile, nonreactive chemical will be invariant as a function of I. Thus, with this transformation, (4.69) becomes

$$\begin{aligned} \mathrm{RMF} &= \frac{J_w}{R}\int_0^\infty \exp(-\mu\, t)\, f^f_m\Big(z, \frac{J_w\, t}{R}\Big)\, dt \\ &= \int_0^\infty \exp\Big(-\frac{\mu\, R\, I}{J_w}\Big)\, f^f_m(z,I)\, dI\;, \end{aligned} \tag{4.70}$$

where $f^f_m(z,I)$ is the normalized travel time pdf in terms of I, which is assumed to be independent of J_w.

Equation (4.70) was used by Jury and Gruber (1989) as a screening model to study the effect of soil and climatic variability on the leaching of pesticide residues below the biologically active zone of soil. Each pesticide is represented in the model through its environmental fate properties (adsorption and decay). To make the adsorption process more independent of a particular soil, the authors divided the distribution coefficient K_d in (4.10) into

$$K_d = f_{oc} K_{oc} \ , \tag{4.71}$$

where f_{oc} is the organic C fraction and K_{oc} is the organic C partition coefficient. This division has been found to reduce significantly the coefficient of variation of adsorption of a given nonionic pesticide between soils (Hamaker and Thompson, 1972). Thus, K_{oc} represents the inherent adsorption potential of a pesticide, and is coupled to a given soil through (4.71).

An illustration of how (4.70) can be used as a screening model to study the effect of soil and environmental conditions is shown in the next example.

Example 4.9 SPATIAL VARIABILITY AND PESTICIDE GROUND WATER POLLUTION POTENTIAL

Equation (4.70) may be used to generate a value for the RMF of a compound under a specific set of soil and environmental conditions once a model is specified for the inherent mobile travel time pdf $f_m^f(z, I)$. The simplest model for a soil with a biologically active zone of thickness ℓ, and a constant water content θ, is the piston flow model.

Case 1: Piston Flow of Solute in Homogeneous Soil
The travel time pdf of the piston flow model in terms of I is (see Example 2.1)

$$f_m^f(\ell, I) = \delta(I - \ell\theta) \ . \tag{4.72}$$

Thus, inserting (4.72) into (4.70), we obtain the corresponding RMF for this case,

$$\mathrm{RMF} = \exp\Big(-\frac{\mu R \ell \theta}{J_w}\Big) \ . \tag{4.73}$$

This formula, which is essentially the model used by Rao et al. (1985) and Jury et al. (1987), can be used as a screening tool for pesticides. Suppose that in a given climate and soil, we require that no pesticide be allowed to have more than some fraction ϵ of its mass penetrate below the surface zone. This requires that RMF $< \epsilon$, or, with (4.73),

$$\exp\Big(-\frac{\mu R \ell \theta}{J_w}\Big) < \epsilon \ . \tag{4.74}$$

If we use the definitions of R and μ, (4.74) may be rewritten as

$$\frac{\ln(2)(\theta + \rho_b f_{oc} K_{oc})\ell}{\tau_{1/2} J_w} > \ln\Big(\frac{1}{\epsilon}\Big) \ . \tag{4.75}$$

Equation (4.75) may be rearranged to form the new expression

$$K_{oc} > a\,\tau_{1/2} - b\ , \tag{4.76}$$

where

$$a = \frac{J_w \ln(1/\epsilon)}{\rho_b f_{oc} \ell \ln(2)}\ , \tag{4.77}$$

$$b = \frac{\theta}{\rho_b f_{oc}}\ . \tag{4.78}$$

Equation (4.76) expresses the condition under which a pesticide with particular values of K_{oc} and τ will leach less than a specified fraction ϵ of the applied mass. The parameters a and b are functions of the soil $(\rho_b, f_{oc}, \theta, \ell)$ and climate (J_w). Figure 4.6 shows a phase space plot of (4.76) for two contrasting situations, representing relatively high and low potential for leaching.

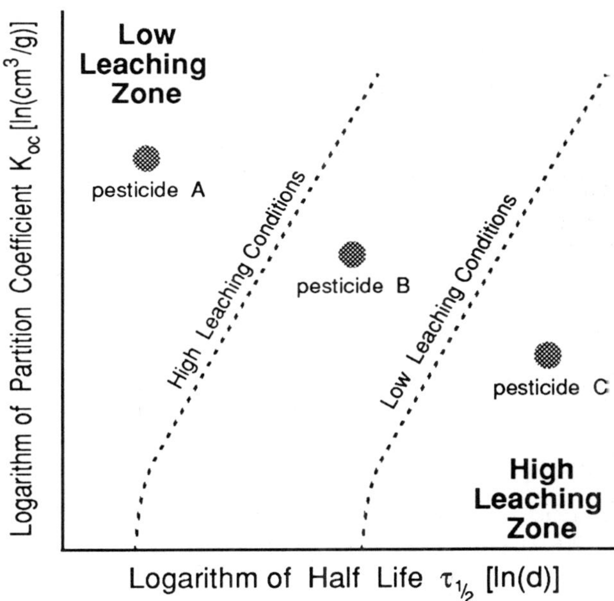

Figure 4.6: Phase space plot of leaching potential for the piston flow screening model. Each pesticide appears as a point in space, and each leaching scenario as a line. Pesticides leaching less than the specified fraction of material are to the left of the line. (Adapted from Jury et al., 1987.)

In this example, pesticide A will have a negligible RMF under both conditions, pesticide C under neither, and the RMF of pesticide B will be negligible under the low condition but not under the high one.

We may examine the effect of soil travel time variabilty on the RMF by allowing $f_m^f(z, I)$ to have a range of values.

Case 2: Variable Soil

In this case, if we let the fundamental soil pdf be described by a gamma distribution (2.61),

$$f_m^f(z, I) = \frac{\beta^{1+\alpha} I^\alpha \exp(-\beta I)}{\alpha!}\ , \tag{4.79}$$

where N and β are constants, then the integral (4.70) can be evaluated analytically. To make the mean transport in this soil agree with the piston flow soil (4.72), we require that (see Problem 2.8)

$$\mathrm{E}(I) = \frac{1+\alpha}{\beta} = \ell\theta \ . \tag{4.80}$$

If we plug (4.79) into (4.70), we obtain, using (D.39)

$$\mathrm{RMF} = \left(1 + \frac{\mu R}{\beta J_w}\right)^{-(1+\alpha)} . \tag{4.81}$$

Using (4.80), (4.81) may be written as

$$\mathrm{RMF} = \left(1 + \frac{\mu R\ell\theta}{(1+\alpha)J_w}\right)^{-(1+\alpha)} . \tag{4.82}$$

4.9.2 Estimation of the Variance of the RMF

There is another way to interpret the RMF (4.69). It is the ensemble average of the local RMF $= \exp(-\mu t_a)$ of a stream tube that has a travel time t_a, integrated over the distribution of adsorbed chemical travel times

$$f_a^f(z, t_a) = \frac{1}{R}\, f_m^f\left(z, \frac{t_a}{R}\right) . \tag{4.83}$$

Thus, the local RMF has a variance given by

$$\mathrm{Var}(\exp(-\mu t)) = \mathrm{E}(\exp(-2\mu t)) - \mathrm{E}^2(\exp(-\mu t)) \ . \tag{4.84}$$

Therefore, using (4.82), we may write the variance as

$$\mathrm{Var}(\exp(-\mu t)) = \left(1 + \frac{2\mu R\ell\theta}{(1+\alpha)J_w}\right)^{-(1+\alpha)} - \left(1 + \frac{\mu R\ell\theta}{(1+\alpha)J_w}\right)^{-2(1+\alpha)} . \tag{4.85}$$

For example, if a pesticide in a given soil and climate situation had $\mu R\ell\theta/J_w = 2$, and the variable soil had a net applied water pdf (4.79) with $\alpha = 2$ (corresponding to a CV of 0.58) then the two models would make the following predictions for the value of the RMF

Piston Flow	RMF $= 0.135$
Variable soil	RMF $= 0.216 \pm 0.179$,

where 0.179 is one standard deviation from the mean calculated with (4.82) and (4.85). Thus, by treating the transport problem as a stochastic stream tube model, uncertainty can be incorporated into the estimate of the RMF.

This model is extended further to cover stochastic water input in section 6.3.

Problems[5]

Problem 4.1 Assume that the soil consists of a set of noninteracting stream tubes and that each tube is characterized by a single solute velocity V and decay constant μ. If each stream tube obeys the piston flow equation

$$\frac{\partial C}{\partial t} + V\frac{\partial C}{\partial z} + \mu C = 0 \;, \tag{4.86}$$

a) Calculate the travel time flux pdf of the stream tube.

b) Calculate the travel time pdf for the transport volume assuming that V is constant and that μ is gamma distributed

$$f_\mu(\mu) = \frac{\beta^{1+\alpha}\mu^N \exp(-\beta\mu)}{N!} \;. \tag{4.87}$$

Does this outflow concentration behave like it has been subjected to first order decay?

c) Calculate travel time pdf for the transport volume assuming that μ is constant and that V is distributed according to the pdf $f_V(V)$. Does this outflow concentration behave like it has been subjected to first order decay?

Problem 4.2 Prove that the travel time pdf of (4.34) is lognormal if $f(R, t_m)$ is bivariate lognormal.

Problem 4.3 In the classic field experiment of Nielsen et al., (1973), the distribution of infiltration rate over a 150 ha field under ponding was measured and found to be lognormal ($\mu_i = 2.58$, $\sigma_i = 1.00$). At the same field, Biggar and Nielsen (1976) studied solute transport under ponded infiltration and found that the apparent travel time[6] of the solute pulse to $\ell = 100$ cm was also lognormally distributed over the field ($\mu_t = -0.58$, $\sigma_t^2 = 1.56$)

Using the assumption that the infiltration rate and inherent soil variability are perfectly correlated during ponding, calculate the parameters of the soil net applied water pdf $f^f(\ell, I)$.

[5] Problems marked with a † are more difficult.

[6] We calculated the travel time pdf from their velocity distribution using (4.4).

Problem 4.4 Prove that if X is lognormally distributed (μ_X, σ_X) then so is $Y = aX^b$, with $\mu_Y = \mu_X + \ln(a)$, $\sigma_Y = b\sigma_X$.

Problem 4.5 Prove equation (4.60).

Problem 4.6 In the field experiment discussed in Example 4.7, assume that the narrow pulse of solute was added at a constant concentration C_0 through the sprinkler over a period of time Δt. Show that the mass added to the field is also spatially variable, and that this can contribute to the variance of the field-averaged travel time pdf.

Problem 4.7 Assume that the field consists of a network of parallel soil columns through which solute flows in accordance to the local CDE, and that the parameters V, D of the local CDE are independent and lognormally distributed over the field. Using the method of moments, calculate the mean and variance of the mean travel time pdf in terms of the parameters of the V and D distributions. Calculate the macrodispersion coefficient of the field-mean process.

†**Problem 4.8** Calculate the zero and first depth moments of the rate limited CDE adsorption model in infinite soil—Example 4.4 extended to infinite soil—assuming an initial condition $C_t^r(z, 0) = \delta(z)$. Calculate expressions for the solute velocity of a mobile and adsorbing chemical as a function of time. Assume that the adsorbed and mobile phases are initially at equilibrium.

Problem 4.9 Derive the travel time mean and variance expressions (4.31)–(4.32) of the CDE model under rate limited adsorption.

Chapter 5

Transfer Functions for Vertically Heterogeneous Soils

In the applications of the transfer function thus far, the process models used to extend the flux pdf to different depths have assumed that the soil is macroscopically homogeneous in the vertical direction. This assumption was made explicitly in Chapter 2 when the stochastic-convective process was defined (2.65), and it is implicit in the convective-dispersive models that have constant parameters which do not depend on z.

However, most soils vary in texture or structure in the z-direction, and some have sharp horizons where their transport and retention properties change abruptly. A transfer function can be defined to characterize the outflow flux concentration from a transport volume which includes heterogeneous soil layers in exactly the same way as for homogeneous soil, because the flux pdf characterizes the travel time distribution through the volume regardless of its internal composition. We cannot extrapolate from one location to another in heterogeneous soil, however, without building a description of the dependence of the soil properties into the process model, or into the flux pdf at different locations. Thus, model assumptions such as the stochastic-convective hypothesis (2.65) will only be valid if the soil is homogeneous.

5.1 Depth-Dependent Water Content

One way in which the soil can be heterogeneous is if the soil water content varies as a function of z. All of the process models we have discussed thus far assume that the travel time under steady water flow increases linearly with depth in homogeneous soils with a uniform water content. Therefore, the simplest way to modify a transfer function to include heterogeneity is to adjust the mean travel time for nonuniformities in the water content distribution, while at the same time investigating the conditions under which the new model will have depth-independent parameters in the new coordinate frame.

The next example, taken from Simmons (1986b), illustrates how this transformation is achieved for the CDE.

Example 5.1 STEADY STATE CONVECTIVE-DISPERSIVE FLOW IN HETEROGENEOUS SOIL

The equation for the CDE flux concentration in heterogeneous soil describing one dimensional transport of solute under steady-state water flow through a soil with a water content $\theta(z)$ is (Simmons, 1986b)

$$\theta(z)\frac{\partial C^f}{\partial t} + J_w\frac{\partial C^f}{\partial z} - \frac{\partial}{\partial z}\left(\theta(z)D(z)\frac{\partial C^f}{\partial z}\right) = 0\ , \tag{5.1}$$

where it is assumed that the dispersion coefficient $D(z)$ may also depend on z. Although in general (5.1) will require a numerical solution, it may be solved analytically for the following special case. First, the depth coordinate z is transformed to the fluid coordinate frame by the operation

$$y := \int_0^z \theta(z')\,dz'\ . \tag{5.2}$$

Equation (5.1) can now be written in terms of y using (5.2) and the chain rule of differentiation (Kaplan, 1984)

$$\frac{\partial}{\partial z} = \frac{dy}{dz}\frac{\partial}{\partial y} = \theta(z)\frac{\partial}{\partial y}\ . \tag{5.3}$$

Thus, after using (5.3) and dividing by $\theta(z)$, we may write (5.1) as

$$\frac{\partial C^f}{\partial t} + J_w\frac{\partial C^f}{\partial y} - \frac{\partial}{\partial y}\left(\theta^2(z)D(z)\frac{\partial C^f}{\partial y}\right) = 0\ . \tag{5.4}$$

Equation (5.4) is now a CDE which has a constant solute velocity in the fluid coordinate frame, and a depth-dependent effective dispersion coefficient. However, in the special case that[1]

$$\theta^2(z)D(z) =: E = \text{constant}, \tag{5.5}$$

(5.4) reduces to

$$\frac{\partial C^f}{\partial t} + J_w\frac{\partial C^f}{\partial y} - E\frac{\partial^2 C^f}{\partial y^2} = 0\ , \tag{5.6}$$

which is a CDE with constant coefficients. Thus, if a narrow pulse of solute is added to the inlet end $z = y = 0$, the outflow flux concentration at any depth $y(z)$ is the travel time pdf for this model. Since (5.6) now has constant coefficients and the same boundary and initial conditions as the CDE in homogeneous soil, the flux pdf is merely the Fickian pdf of the homogeneous CDE (2.51) with the appropriate change of variables

$$z \longrightarrow y(z)\ , \qquad V \longrightarrow J_w\ , \qquad D \longrightarrow E\ . \tag{5.7}$$

[1] We are aware of no theoretical justification for this relationship between D and θ.

Therefore, in the normal coordinate frame the flux pdf is

$$f^f(z,t) = \frac{y(z)}{2\sqrt{\pi E t^3}} \exp\Big(-\frac{\big(y(z) - J_w t\big)^2}{4Et}\Big) . \tag{5.8}$$

where $y(z)$ is given by (5.2).

The flux pdf (5.8) has a characteristic skewed shape which does not betray the presence of heterogeneity when viewed at a given location z, because y is a constant. However, the resident pdf corresponding to this travel time pdf is (see Problem 5.1)

$$\begin{aligned} f^r(z,t) &= \frac{\theta(z)}{\sqrt{\pi E t}} \exp\Big(-\frac{\big(y(z) - J_w t\big)^2}{4Et}\Big) \\ &\quad - \frac{J_w \theta(z)}{2E} \exp\Big(\frac{J_w y(z)}{E}\Big) \operatorname{erfc}\Big(\frac{y(z) + J_w t}{\sqrt{4Et}}\Big) , \end{aligned} \tag{5.9}$$

which clearly will reveal the presence of heterogeneity since it is a function of z. Figure 5.1 shows an illustration of (5.9) in a hypothetical soil which has a water content which varies sinusoidally with depth z,

$$\theta(z) = \theta_0 + \theta_1 \sin(\lambda z) . \tag{5.10}$$

The resident pulse, which is Gaussian in the fluid coordinate y-frame, has a strange shape in the normal coordinate system, reflecting the spatial dependence of the water content.

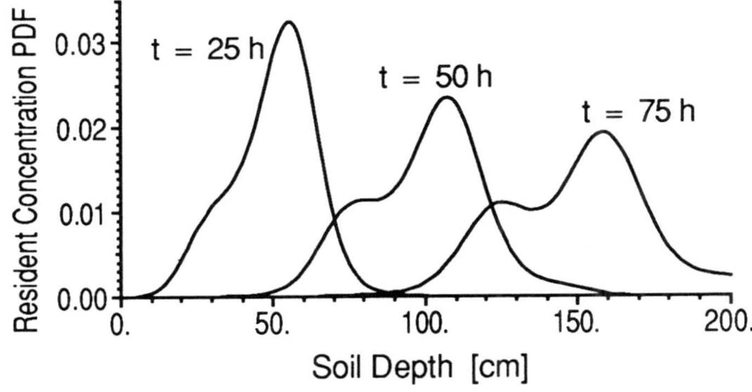

Figure 5.1: *Resident concentration pdfs for the heterogeneous CDE (5.9) with* $J_w = 1$ *cm/hr and* $E = 1$ *cm*^{2}*/hr, in a soil with a sinusoidal water content profile (5.10) with* $/th_0 = 0.5$, $\theta_1 = 0.15$, *and* $\lambda = 0.125$ *cm*$^{-1}$.

Although this manner of treating heterogeneity has made a rather restrictive *ad hoc* assumption about the water content dependence of dispersion ((5.5) must be valid), it was shown to remove virtually all of the heterogeneity of the CDE transport coefficients measured in large field plots at various locations on a 1.4 ha field (Ellsworth, 1989; Ellsworth et al. 1990). In this study, the authors were able to describe the area-averaged movement of discrete solute plumes added as massive pulses to square plots 1.5 m on each side in the fluid coordinate frame $y(z)$, using single values of J_w and E for the entire field.

We may also modify the stochastic-convective hypothesis (2.65) in a heterogeneous system in a similar manner by switching to a fluid coordinate frame. This produces (see Problem 5.2)

$$f^f(z,t) = \frac{y(\ell)}{y(z)} f^f\Big(\ell, \frac{ty(\ell)}{y(z)}\Big) , \tag{5.11}$$

where ℓ is the reference depth and $y(z)$ is defined by (5.2). Equation (5.11) reduces to (2.65) when θ is constant. This model has not been tested in heterogeneous soils.

The simple approach discussed above was motivated more by the desire for mathematical simplicity than by physical insight. An alternative approach to describing transport through heterogeneous soil is to divide the transport volume into a series of layers, each of which is characterized by its own travel time pdf. This method is discussed in the next section.

5.2 Solute Transport in Layered Soils

In contrast to the situation in ground water, the normal direction of flow in the vadose zone is perpendicular to the natural layering of the soil. This section will discuss how to formulate the problem of solute transport through a series of soil layers of different properties as a transfer function. The two layer problem will be described in detail, because it is easily generalized to N layers.

Figure 5.2 shows an idealized picture of solute transport through a two layer soil profile made up of two layers of contrasting properties.

The soil consists of two layers, each of which is described by a travel time pdf. The travel time pdf $f_1^f(L_1, t_1)$ of layer 1 could be measured by adding a narrow pulse of solute to the inlet end, and measuring the outflow concentration at $z = L_1$ as a function of time. However, the travel time pdf $f_2^f(L_2, t_2)$ of layer 2 normally cannot be measured directly, since it is not possible to add a narrow pulse of solute to the inlet end of the second layer at $z = L_1$. The measurement of the outflow flux concentration at $z = L_1 + L_2$, in response to a narrow pulse input at $z = 0$, can be used to construct the travel time pdf $f^f(z,t)$, where t

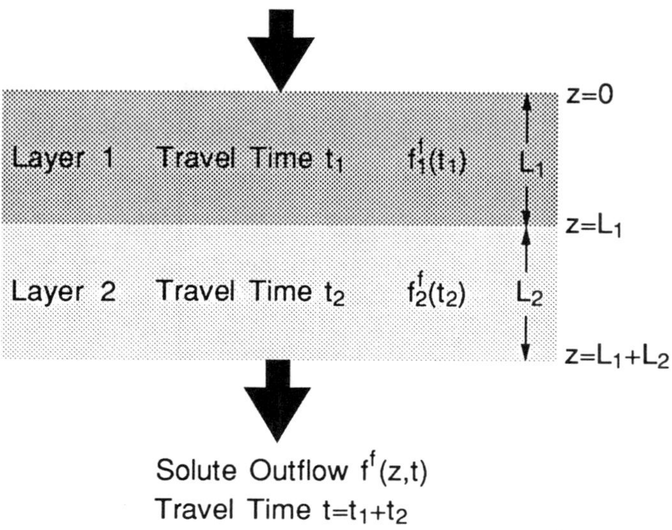

Figure 5.2: *Transfer function description of transport through layered soil. The travel time required to reach the outflow end is the sum of the travel times required to move through the two layers, each of which is described by a travel time pdf.*

is the sum of the travel times through layers 1 and 2. Thus, the travel time t must satisfy the constraint that $t = t_1 + t_2$.

The travel time pdf at depth z is therefore given by

$$f^f(z,t) = \int_0^t \int_0^{t-t_2} \delta(t - t_1 - t_2)\, f_{12}(t_1, t_2)\, dt_1 dt_2 \ , \tag{5.12}$$

where $f_{12}(t_1, t_2)$ is the joint pdf of t_1 and t_2. Equation (5.12) may be thought of as a stochastic stream tube model where the local stream tube model is $\delta(t - t_1 - t_2)$. Thus, the outflow at z will depend in general on the travel time distribution of each layer, and on the correlations between the travel times of different layers. There are two special cases of interest which are discussed below, each of which will lead to simplified solutions of (5.12). First, however it is useful to review some basic properties of random variables.

Excursus Sums and Products of Random Variables

If X_1 and X_2 are random variables which are described by a joint pdf $f(X_1, X_2)$, then the random variable $Z = X_1 + X_2$ has the following moments (Himmelblau, 1970):

$$\mathrm{E}(Z) = \int_{-\infty}^{\infty} \int_{-\infty}^{\infty} (X_1 + X_2)\, f(X_1, X_2)\, dX_1 dX_2 = \mathrm{E}(X_1) + \mathrm{E}(X_2) \ , \tag{5.13}$$

$$\mathrm{Var}(Z) = \mathrm{Var}(X_1) + \mathrm{Var}(X_2) + 2\,\mathrm{Cov}(X_1, X_2) \;, \tag{5.14}$$

where

$$\begin{aligned} \mathrm{Cov}(X_1, X_2) &:= E\Big[\big(X_1 - \mathrm{E}(X_1)\big)\big(X_2 - \mathrm{E}(X_2)\big)\Big] \\ &= \mathrm{E}(X_1 X_2) - \mathrm{E}(X_1)\,\mathrm{E}(X_2) \end{aligned} \tag{5.15}$$

is the covariance of X_1 and X_2. The covariance is related to the correlation coefficient ρ by

$$\rho := \frac{\mathrm{Cov}(X_1, X_2)}{\sqrt{\mathrm{Var}(X_1)\,\mathrm{Var}(X_2)}} \;. \tag{5.16}$$

It follows from (5.15) that the random variable defined by the product $X_1 X_2$ has an expectation, or ensemble average, given by

$$\mathrm{E}(X_1 X_2) = \mathrm{E}(X_1)\,\mathrm{E}(X_2) + \mathrm{Cov}(X_1, X_2) \;. \tag{5.17}$$

Using (5.16), this may be expressed as

$$\mathrm{E}(X_1 X_2) = \mathrm{E}(X_1)\,\mathrm{E}(X_2) + \rho\sqrt{\mathrm{Var}(X_1)\,\mathrm{Var}(X_2)} \;. \tag{5.18}$$

These relationships may be extended to the case of the sum

$$Z := \sum_{i=1}^{N} X_i \tag{5.19}$$

of N random variables X_i, $i = 1, \ldots, N$. If we denote the mean and variance of X_i by μ_i and σ_i, then it can be shown that the mean and variance of Z are (see Problem 5.3)

$$\mathrm{E}(Z) = \sum_{i=1}^{N} \mathrm{E}(X_i) = \sum_{i=1}^{N} \mu_i \;, \tag{5.20}$$

$$\mathrm{Var}(Z) = \sum_{i=1}^{N}\sum_{j=1}^{N} \rho_{ij}\sigma_i\sigma_j \;, \tag{5.21}$$

where ρ_{ij} is the correlation coefficient of X_i and X_j.

In general, as seen by (5.21), the travel time pdf through a series of soil layers will depend on the correlation structure of the travel times in adjacent layers. However, for the special cases of independence (where $\rho = 0$) and perfect correlation ($\rho = 1$), the travel time pdf through the system may be expressed completely in terms of the properties of the individual layers. These cases are described below.

5.2.1 Independent Soil Layers

If the travel times through the two soil layers are independent of each other, then the joint pdf may be written as

$$f_{12}(t_1, t_2) = f_1^f(L_1, t_1) f_2^f(L_2, t_2) \; . \tag{5.22}$$

For this case, after integrating over t_2, (5.12) may be written as

$$f^f(L_1 + L_2, t) = \int_0^t f_1^f(L_1, t_1)\, f_2^f(L_2, t - t_1)\, dt_1 \; . \tag{5.23}$$

Equation (5.23) has a particularly simple form after Laplace transformation (see (A.36) of Appendix A)

$$\hat{f}^f(L_1 + L_2; s) = \hat{f}_1^f(L_1; s) \hat{f}_2^f(L_2; s) \; . \tag{5.24}$$

The assumption of independent travel times between soil layers might be appropriate for certain types of problem, such as the flow of solute from the soil surface to a tile drain. In that system, the distribution of travel times through the unsaturated zone would be affected primarily by the texture and structure of the soil, whereas the travel time through saturated soil is influenced significantly by the position of the point of entry at the water table relative to the position of the drain tubes (Jury, 1975). Utermann et al. (1990) recently modeled solute transport through a tile drained field with a transfer function using (5.23).

5.2.2 Perfectly Correlated Soil Layers

If the travel times through the second soil layer are perfectly correlated[2] with the travel times of the first layer, then the travel time t_2 is completely specified once t_1 is known. This condition is expressed formally as

$$f_{12}(t_1, t_2) = f^f(L_1, t_1) \delta(t_2 - g(t_1)) \; , \tag{5.25}$$

where $t_2 = g(t_1)$ expresses some functional dependence between t_2 and t_1. Once $g(t_1)$ has been specified, the travel time pdf can be calculated by inserting (5.25) into (5.12) (see Problem 5.4). The exact nature of the functional relation between t_2 and t_1 will depend on how they are distributed. Since we wish to represent the case where the relative order of travel times are the same in

[2] This does not necessarily imply linear correlation.

each layer, the function $t_2 = g(t_1)$ is derived by equating the cdfs of the two distributions

$$P_1(L_1, t_1) = P_2(L_2, t_2) \ . \tag{5.26}$$

Equation (5.26) has the formal solution

$$t_2 = P_2^{-1}(P_1(L_1, t_1)) \ . \tag{5.27}$$

For example, if t_1 and t_2 are both normally distributed and perfectly correlated, then (5.26) reduces to

$$\frac{t_1 - m_1}{s_1} = \frac{t_2 - m_2}{s_2}, \tag{5.28}$$

or,

$$t_2 = m_2 - m_1\frac{s_2}{s_1} + t_1\frac{s_2}{s_1} = \alpha + \beta t_1 \ , \tag{5.29}$$

where m_i and s_i^2 are the mean and variance of the travel time in layer i. Similarly, if each layer has lognormally distributed travel times, then (5.26) implies that (see Problem 5.9)

$$\frac{\ln(t_1) - \mu_1}{\sigma_1} = \frac{\ln(t_2) - \mu_2}{\sigma_2}, \tag{5.30}$$

which produces a nonlinear dependence between t_1 and t_2

$$t_2 = \exp\Big(\mu_2 - \mu_1\frac{\sigma_2}{\sigma_1}\Big)\, t_1^{\sigma_2/\sigma_1} = \alpha t_1^{\beta} \ , \tag{5.31}$$

where μ and σ^2 are the mean and variance of $\ln(t)$. An illustration of how to calculate the travel time pdf (5.12) through two soil layers whose travel times are perfectly correlated and lognormal is given in Problem 5.4.

An example of a situation where one might expect the travel times of adjacent soil layers to remain highly correlated is one in which the layers are all highly permeable, and the variation in travel times within a layer is caused primarily by local variations in water flux. In such a case, the short travel times through layer 1, which are caused by a high local water flux, become the short travel times in layer 2, as long as the water flux is continuous. Similarly, the long travel time regions of layer 1 which have low water flux will become the long travel time regions of layer 2 (see Figure 5.3).

The convective-dispersive and stochastic-convective models have been singled out for discussion in earlier chapters because they represent different extremes for solute dispersion. In addition, they represent extremes in solute

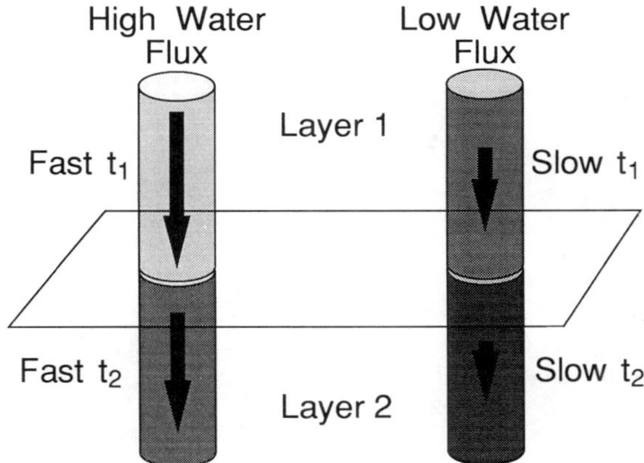

Figure 5.3: Schematic diagram of two stream tubes containing different local water fluxes, giving rise to correlated travel times in adjacent soil layers, as long as the water fluxes remain continuous locally (i.e. no lateral flow).

travel time correlation as well, with the convective-dispersive model representing a soil with uncorrelated travel times and the stochastic-convective model representing one with perfectly correlated travel times. This statement is proven in the next two examples.

Example 5.3 THE ZERO-CORRELATION CONVECTIVE-DISPERSIVE MODEL

The travel time to depth z represented by the flux pdf of the CDE (2.51) in homogeneous soil has a mean and variance given by (2.57)–(2.58). Even though the soil is homogeneous in the vertical direction, we may think of it as a series of N layers of thickness Δz, each of which has an identical travel time pdf. This pdf is also described by the Fickian model (2.51), and has a mean and variance given by (2.57)–(2.58), with $z \to \Delta z$. At a depth $z = N\Delta z$ we may write the mean and variance as

$$\mathrm{E}_z(t) = \sum_{j=1}^{N} \mathrm{E}(t_j) = \frac{z}{V} = N\frac{\Delta z}{V} = N\,\mathrm{E}_{\Delta z}(t_1)\ , \tag{5.32}$$

$$\mathrm{Var}_z(t) = \frac{2Dz}{V^3} = N\left(\frac{2D\Delta z}{V^3}\right) = N\,\mathrm{Var}_{\Delta z}(t_1)\ . \tag{5.33}$$

Since (5.32)–(5.33) are valid by assumption at every z, the CDE is a model whose travel time has a mean and variance equal to the sum of the means and variances of the individual layers. Thus, by (5.21), the correlation coefficient ρ_{ij} between different layers $i \neq j$ must be zero.

In Chapter 2, we defined a stochastic-convective model as one whose travel time pdf obeyed (2.65) in homogeneous soil. The mean and variance of the travel time of this process are given in (2.67)–(2.68). Therefore, unlike the CDE, the travel-time variance increases as the square of the distance beyond the inlet end in a stochastic-convective model. This property is equivalent to perfect correlation, as shown in the next example.

Example 5.4 THE PERFECT CORRELATION STOCHASTIC-CONVECTIVE MODEL

If N random variables $t_1, \ldots, t_N$ are identically distributed and perfectly correlated, then, since $\rho_{ij} = 1$, (5.20)–(5.21) become

$$\mathrm{E}_z(t) = \sum_{i=1}^{N} \mathrm{E}_{\Delta z}(t_i) = \sum_{i=1}^{N} \mathrm{E}_{\Delta z}(t_1) = N\,\mathrm{E}_{\Delta z}(t_1)\ , \tag{5.34}$$

$$\mathrm{Var}_z(t) = \sum_{i=1}^{N}\sum_{j=1}^{N} \mathrm{Var}_{\Delta z}(t_1) = N^2\,\mathrm{Var}_{\Delta z}(t_1) = \left(\frac{z}{\Delta z}\right)^2 \mathrm{Var}_{\Delta z}(t_1)\ , \tag{5.35}$$

which is identical to (2.67)–(2.68). Thus, stochastic-convective models are perfect correlation models.

5.3 Flux PDFs in Layered Soils

Correlation between travel times of adjacent soil layers can clearly have a significant influence on the transport process in layered soils. However, the shape of the flux pdf at the outlet end of a soil made up of two layers of different properties does not change significantly as the correlation coefficient between the two layers is changed. Figure 5.4 shows two travel time pdfs of 30 cm soil

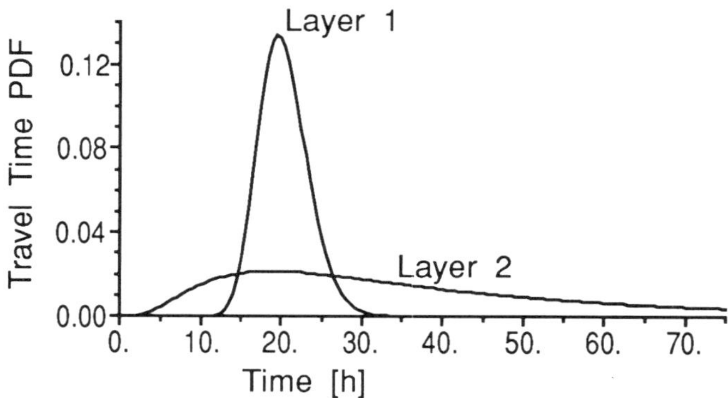

Figure 5.4: Travel time pdfs for 30 cm soil layers through which water is flowing at a steady flux of 1 cm/hr. The pdfs are represented equally well as CLT (2.72) or CDE (2.51) models with parameters $\mu_1 = 3.0$, $\sigma_1 = 0.15$, $V_1 = 1.48$, $D_1 = 0.5$ and $\mu_2 = 3.5$, $\sigma_2 = 0.75$, $V_2 = 0.68$, $D_2 = 7.7$, respectively, for layers 1 and 2.

layers which have significantly different transport properties, and Figure 5.5 shows the predicted flux pdfs at the 60 cm depth—using (5.12) and assuming

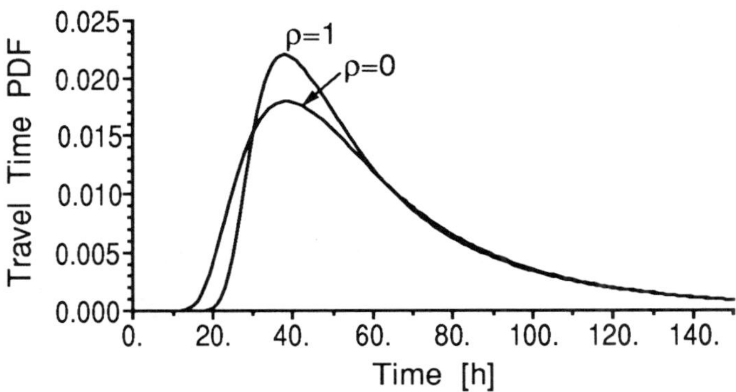

Figure 5.5: *Travel time pdfs at $z = L_1 + L_2 = 60$ cm for the soil with properties shown in Fig. 5.4, calculated assuming zero and perfect correlation between the layers.*

that the two layers are uncorrelated ($\rho = 0$) or perfectly correlated ($\rho = 1$).

This result indicates that correlation structure would be difficult to identify from a flux pdf measurement through a layered soil, even if the travel time pdfs of each layer were known. More common is the case where only the pdf of layer 1 is known, so that both the pdf of layer 2 and the correlation coefficient would have to be fitted to the observation of the flux pdf through the two layers. Moreover, the flux pdfs representing the travel time distribution through layered soil may differ little in appearance from the pdfs of homogeneous soil (Fig. 5.5).

One reason that the flux pdfs do not reveal the presence of heterogeneity is that all solute must pass through each layer before reaching the outflow end. In contrast, the resident pdfs of heterogeneous soil will be quite distinct from those of homogeneous soil, because during transit through the layers, part of the solute pulse will be on each side of the boundary between the layers.

5.4 Convective-Dispersive Resident PDFs for Layered Soil

As we have seen above, development of continuum flux and resident pdfs for layered soil requires specifying both the process model (or travel time pdf) within each layer and the correlation assumptions at the interfaces between layers. Very little work has been done in this area, chiefly because the principal models used today all assume that the soil properties do not vary in the vertical direction. The CDE has been used to model the problem of solute transport through layered soil (Kreft and Zuber, 1978; Barry and Parker, 1987), by assuming that each layer obeys an equation of the form

$$\frac{\partial C_j^r}{\partial t} = D_j \frac{\partial^2 C_j^r}{\partial z^2} - V_j \frac{\partial C_j^r}{\partial z} \tag{5.36}$$

where C_j^r is the resident fluid concentration in layer j $(L_j < z < L_{j+1})$, and V_j, D_j are the solute velocity and dispersion coefficients in this layer. Equation (5.36) is also obeyed by the flux concentration in each layer. Two conditions relating the concentrations in the adjacent zones are needed at the interface between layers. Kreft and Zuber (1978) and Barry and Parker (1987) assumed that the flux and resident fluid concentrations were continuous across the interface. In terms of the resident concentration, this implies that

$$D_j \frac{\partial C_j^r}{\partial z} - V_j C_j^r = D_{j+1} \frac{\partial C_{j+1}^r}{\partial z} - V_{j+1} C_{j+1}^r \ , \tag{5.37}$$

$$C_j^r = C_{j+1}^r \ . \tag{5.38}$$

The corresponding relations for the flux concentration are (Kreft and Zuber, 1978; see also Problem 5.6)

$$C_j^f = C_{j+1}^f \tag{5.39}$$

$$V_j \frac{\partial C_j^f}{\partial z} = V_{j+1} \frac{\partial C_{j+1}^f}{\partial z} \tag{5.40}$$

If the solute travel time pdf at $z = L_1 + L_2$ is calculated in the two layer soil by solving (5.36)–(5.40) and the travel time moments are analysed, one finds that, in contrast to the result in homogeneous soil (Example 5.3), the variance at $z = L_1 + L_2$ is not equal to the sum of the variances in the individual layers (Barry and Parker, 1987). Therefore, by (5.14), the covariance or correlation coefficient between the two layers must be nonzero. Thus, the CDE, whose travel times are uncorrelated within a homogeneous profile, has correlated travel times between adjacent regions in layered soil. Since the assumption of flux continuity (5.37) is obviously valid, the reason for this strange result must be the assumption that the resident fluid concentration is continuous at the interface in (5.38) or (5.40).

We can gain further insight into what is happening to the resident concentration at the interface by solving for the concentration distribution in a system where each layer obeys the CDE (5.36), but where the travel times of the two layers are assumed to be uncorrelated. For this system we use the layered soil travel time model for independent layers (5.23), with each layer represented by the Fickian travel time pdf (2.51). Provided that we can generate an expression for the flux concentration at every location z in the profile, we can use (3.5) to calculate the corresponding resident concentrations. The travel time pdf of the CDE with independent layers may be written as

$$f^f(z,t) = \begin{cases} f_1^f(z,t) & , 0 < z \leq L_1 \; ; \\ \displaystyle\int_0^\infty f_1^f(L_1,t_1)\, f_2^f(z-L_1,t-t_1)\, dt_1 & , z > L_1 \; . \end{cases} \tag{5.41}$$

Note that the flux pdf is simply equal to the flux pdf of layer 1 at depth z when $z < L_1$, because the soil is homogeneous to this depth. For $z > L_1$, (5.23) is used to represent transport through a layered soil whose upper and lower layers have thickness L_1, and $z - L_1$, respectively.

The Laplace transform solutions of (5.41) are, using (5.24)

$$\widehat{f}^f(z;s) = \begin{cases} \widehat{f}_1^f(z;s) & , 0 < z \leq L_1 \; ; \\ \widehat{f}_1^f(L_1;s)\, \widehat{f}_2^f(z-L_1;s) & , z > L_1 \; . \end{cases} \tag{5.42}$$

The resident fluid concentrations corresponding to the flux concentrations in (5.42) may be calculated with (3.7)

$$\widehat{C}_l^r(z;s) = \begin{cases} -\dfrac{V_1}{s}\dfrac{d\widehat{f}_1^f(z;s)}{dz} & , 0 < z \leq L_1 \; ; \\ -\dfrac{V_2}{s}\widehat{f}_1^f(L_1;s)\dfrac{d\widehat{f}_2^f(z-L_1;s)}{dz} & , z > L_1 \; , \end{cases} \tag{5.43}$$

where we have made the substitution

$$\widehat{C}_l^r(z;s) = \begin{cases} \dfrac{\widehat{C}_t^r(z;s)}{\theta_1} & , 0 < z \leq L_1 \; ; \\ \dfrac{\widehat{C}_t^r(z;s)}{\theta_2} & , z > L_1 \; . \end{cases} \tag{5.44}$$

The two pdfs in (5.43) are represented by the Laplace transform of the Fickian pdf (2.50) with the appropriate values of z, V, and D. The concentrations may be obtained as a function of z and t by inverting the Laplace transforms numerically (see Appendix B).

Figure 5.6 shows the resident fluid concentration profiles predicted with the two layer CDE assuming continuous resident concentration (5.36)–(5.38) and assuming independent layers (5.43) for a layered soil with short, uniform travel

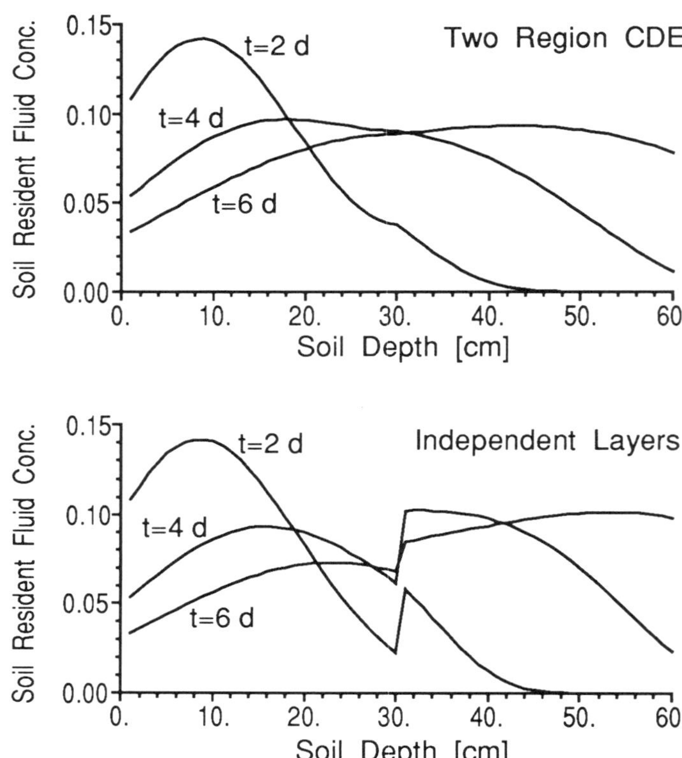

Figure 5.6: *Resident fluid concentrations for the two layer CDE calculated assuming continuous resident concentration and independent layers for a soil with $V_2 = 3$ cm/d, $D_2 = 35$ cm^2/d in the lower 30 cm, and $V_1 = 11$ cm/d, $D_1 = 3$ cm^2/d in the upper 30 cm.*

times in the upper layer, and long, highly variable travel times in the lower layer.

In contrast to the flux concentration (Fig. 5.5), the resident concentration shape (Fig. 5.6) is affected significantly by the interface assumptions. The transport processes are represented in an identical manner in each region, but the assumptions made about the interface have produced very different features in the profiles of the two models. The discontinuity in the resident concentration calculated when the layers are assumed to be independent is caused by the pileup of solute in the slower-moving second region. This is avoided in the two-region CDE by altering the profiles in the first region to force continuity at the interface.

Neither model is correct *a priori*. However, since the CDE model has uncorrelated travel times in homogeneous soil, is seems strange that a correlation would develop as the interface between two layers that individually obey this hypothesis. Although it seems reasonable that the resident fluid concentration would be continuous locally within a single stream tube, it is not clear that the volume-averaged resident fluid concentration would remain continuous when both concentration and water content might vary laterally in each region.

The differences in predicted profile shape shown in Figure 5.6 are significant enough that the interface hypotheses might be studied experimentally in layered soil transport experiments. However, since the resident fluid concentration is not directly measurable, the predicted shapes of the total resident concentration would have to be compared with observation. Each model would predict a discontinuity of the total resident concentration, but the differences in predicted shape would be comparable to the differences shown in Figure 5.6 (Jury and Utermann, 1991).

5.5 Stochastic-Convective Resident PDFs for Layered Soil

The stochastic-convective model (2.65) was shown to have perfectly correlated travel times in Example 5.4 above. Therefore, it seems reasonable to develop a stochastic-convective model of layered soil by requiring the travel time pdfs to obey (2.65) within a layer, and to be perfectly correlated at the interfaces.

Since the CDE and CLT flux pdfs are nearly identical in shape when fitted to a common observation (see Fig. 2.3), the lognormal model could be used together with (5.31) to represent perfect correlation for the conditions in Figure 5.6. Figure 5.7 shows the resident fluid concentration calculated (see Problems 5.4–5.5) using (5.12), (5.25) and (5.31), assuming that each layer obeys the CLT.

Notable in Figure 5.7 is the much higher peak concentration in the first layer compared to the CDE simulations in Figure 5.6, because the dispersion coefficient of the CLT is increasing linearly in this zone from zero up to a maximum equal to the CDE dispersion coefficient at $z = L$. However, even though these two models look enormously different as resident concentrations, they show only a small difference in outflow shape at the end of the bottom layer (Fig. 5.8).

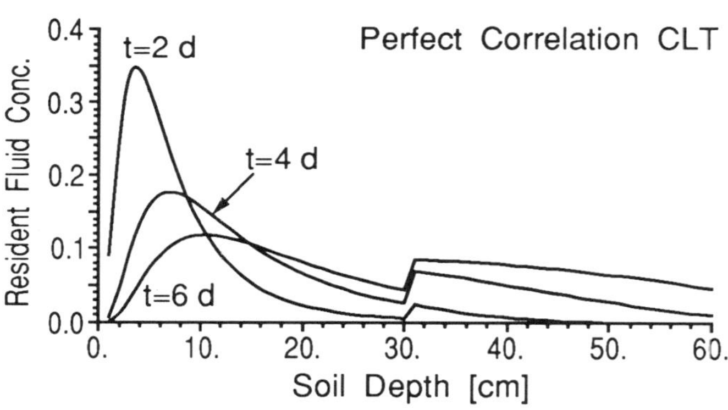

Figure 5.7: Soil resident fluid concentrations corresponding to the conditions in Figures 5.5 and 5.6, except that each layer obeys the CLT and the correlation between layers is assumed to be perfect.

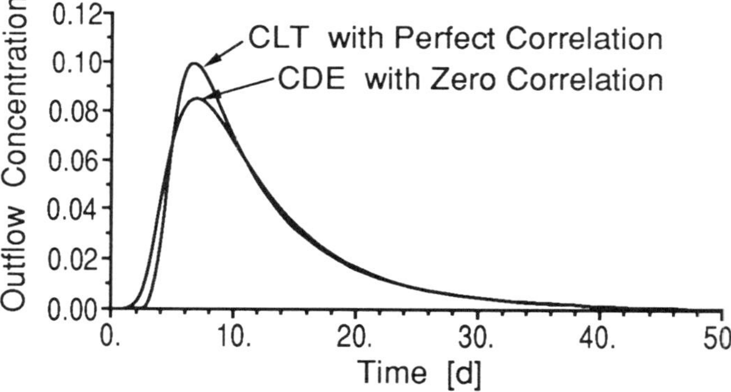

Figure 5.8: Outflow concentrations predicted for the CDE and CLT models, assuming zero and perfect correlation, respectively at the bottom of the second layer $Z = 60$ cm.

It is clear that the formulation of solute transport through layered soil profiles is an unresolved research area. In the future, it will be necessary to conduct both fundamental and applied research studies of transport across interfaces to develop the understanding necessary to advance towards a comprehensive theory. In the interim, it may be necessary to rely on simpler approaches such as the ones outlined above to develop approximate solutions through media with pronounced layering. As a rough rule, it may be possible to approximate the correlation structure with the independent hypothesis whenever pronounced lateral flow occurs at the interfaces, and with perfect correlation whenever all layers are permeable at the highest flow rates. In the event that an instability develops at the interface (Raats, 1973; Hillel, 1986), or when the extent of lateral mixing in the soil shows a pronounced dependence on initial water content (Ellsworth, 1989), a more complex representation than the one outlined above will be necessary to model the resident concentration within the profile.

5.6 Solute Transport Under Depth-Dependent Adsorption

The layered soil approach is also useful for solving transport problems in which the adsorption or degradation rate (see Problem 5.11) is depth dependent. The procedure is illustrated in the next example.

Example 5.5 DEPTH-DEPENDENT ADSORPTION

Assume that a soil is stochastic-convective and homogeneous with a lognormal travel time pdf (2.72) for transport of a mobile chemical, but has a depth-dependent retardation factor given by

$$R(z) = \begin{cases} R_1 & ; 0 < z < \ell \\ R_2 & ; z > \ell \end{cases} . \tag{5.45}$$

Since the soil is stochastic-convective and homogeneous for the mobile chemical, it is reasonable to model the transport for the adsorbing chemical with the perfect correlation layered soil model (5.25). For $z < \ell$, the flux pdf is given by (4.12), with $R = R_1$. For $z > \ell$, the procedure used in Section 5.2.2 is employed to generate the pdf.

Thus, assuming that the lognormal model parameters for the mobile chemical are μ, σ we may write (5.31) at a position $z > \ell$ as

$$\frac{1}{\sigma}\left[\ln\left(\frac{t_2 \ell}{R_2(z-\ell)}\right) - \mu\right] = \frac{1}{\sigma}\left[\ln\left(\frac{t_1}{R_1}\right) - \mu\right] , \tag{5.46}$$

which reduces to

$$t_2 = \frac{z-\ell}{\ell}\,\frac{R_2 t_1}{R_1} = g(t_1) . \tag{5.47}$$

Thus, for $z > \ell$, the flux pdf is obtained by combining (5.47) with (5.25), and inserting the result into (5.12)

$$f^f(z,t) = \int_0^t \delta\left(t - t_1 - \frac{z-\ell}{\ell}\frac{R_2 t_1}{R_1}\right)\frac{1}{R_1} f_m^f\left(\frac{t_1}{R_1}\right) dt_1 . \tag{5.48}$$

This integral may be evaluated by (2.10), with the result

$$f^f(z,t) = \begin{cases} f^f_m(t/R_1)/R_1 & 0 < z < \ell \\ f^f_m(t/\omega(z))/\omega(z) & z > \ell \end{cases}, \tag{5.49}$$

where $\omega(z) = R_1 + R_2(z - \ell)/\ell$.

Problems[3]

Problem 5.1 Derive equation (5.9)

Problem 5.2 Derive Equation (5.11) for the stochastic convective travel time pdf of a heterogeneous soil with a depth-dependent water content.

Problem 5.3 Show that the mean and variance of the sum of N random variables are given by (5.20)–(5.21).

†Problem 5.4 Derive an analytic expression for the travel time pdf at the bottom of two soil layers of thickness ℓ which are bivariate-lognormally distributed and perfectly correlated (5.31).

†Problem 5.5 Calculate the resident pdf for $0 < z < 2\ell$ associated with Problem 5.4.

Problem 5.6 Show that the condition of continuous resident concentration (5.38) is equivalent to the condition (5.40) for flux concentration at the interface of two heterogeneous soil layers.

Problem 5.7 A soil which obeys the stochastic-convective hypothesis is homogeneous except at the depth $Z = \ell$, where the travel times become uncorrelated (i.e. the stochastic-convective regions above and below $Z = \ell$ are uncorrelated). Calculate the travel time variance as a function of depth and show that the mean dispersion coefficient of the system undergoes a decrease after $Z = \ell$.

[3]Problems marked with a † are more difficult.

Problem 5.8 A soil which obeys the stochastic-convective hypothesis is homogeneous except in the region $\ell < Z < 2\ell$, where the mean travel time per unit depth is twice as long as in the rest of the soil volume. However, the layers remain perfectly correlated at the interfaces $Z = \ell, 2\ell$. Calculate the travel time variance as a function of depth and show that the mean dispersion coefficient of the system undergoes a decrease after $Z = \ell$. Assume that the variance per unit length of the soil is constant everywhere.

Problem 5.9 Show that the resident concentration (5.43) for the case of two independent layers with Fickian (CDE) flux pdfs (2.59) is discontinuous at the interface $z = \ell$.

†**Problem 5.10** Calculate the Laplace transform of the solution for the flux concentration of the two-layer CDE (5.36) with the boundary conditions (5.37)–(5.38).

Problem 5.11 Calculate the flux pdf of a mobile solute which undergoes first order decay in the surface zone $(0 < z < \ell)$ with a rate coefficient μ, and does not decay below this depth. Assume that the soil is homogeneous and that a mobile, nondecaying solute has a pdf which obeys the stochastic-convective hypothesis (2.65).

Chapter 6

Practical Field Application of the Transfer Function Approach in Unsaturated Soil

Modelling solute transport through unsaturated soil at the field scale is extremely difficult for a number of reasons. First, lateral and vertical variability of the transport and retention properties that influence water and solute movement cause the flow paths to be extremely complicated. These three-dimensional flow paths are not described in detail by any existing transport models. However, even if transport models could be developed which took such complexities into account at the local scale, they could not be calibrated by any conceivable set of measurements at the field scale. The number of measurements would be prohibitive, and for some properties would require massive disturbance of the soil.

Second, the water flux regime is always transient near the surface, even in locations where water application is frequent and causes net downward flow of solutes. Therefore, solute transport models which are based on the assumption of steady water flow may have to be adjusted to operate under transient flow.

Finally, many unsaturated soil profiles are distinctly heterogeneous along the direction of flow. Abrupt changes in soil properties at soil horizons can cause chaotic flow features to occur, ranging from lateral subsurface flow to local channeling of fluid along a small fraction of the wetted soil volume (Kung, 1988).

Because of the sheer size of the transport volume in many field applications of solute transport, no model can be expected to contain process descriptions that describe all of the potential solute pathways in detail. In any field study or application, the size and quality of the data characterizing the transport and retention properties of the soil will always be a major factor limiting the type of model that can be developed. Models requiring input and calibration data that are far in excess of feasible supplies are not going to be useful, and in the worst case may be misused by requiring many of their parameters to be assigned default values in the absence of measurement.

Transfer functions provide an alternative approach to use in an application where data is difficult to obtain. Although the transfer function model has been developed on a narrow foundation in previous chapters (steady water flow, no spatial correlation structure for hydraulic properties, zero or perfect correlation at layer interfaces), it can be extended considerably by making additional assumptions. This chapter will discuss several of these adaptations, with reference to field testing where appropriate.

6.1 A Transient Transfer Function Model

The field regime is inherently transient. Diurnal cycling of radiation causes temperature and evaporation fluctuations which mediate many processes of interest in solute transport. Water additions are never steady in the field, even in experiments. For these reasons, the solute lifetime or travel time pdf is nonstationary in time[1] and therefore useless; the experimental conditions under which it was measured would never be repeated, so no predictions about future behavior could be made. However, deterministic process models of water and solute movement are impractical for large areas, because of the difficulty of measuring transport properties at a sufficiently high density to allow accurate space and time representation of the local fluxes. To proceed further in the description of transport under these conditions, some additional assumptions must be introduced to simplify the data acquisition process.

Jury et al. (1990) developed a transfer function model for the field regime designed to operate under conditions where solute leaching was likely; i.e. net water flux was downward most of the time. The goal of the modeling exercise was to produce a description of the solute transport process that could be used with a simple model of the soil hydrologic regime, thereby avoiding detailed measurements of the hydrologic and retention properties at a large number of sites.

Their model is based on the following assumptions:

[1] In the language of Chapter 2, this would mean that the travel or lifetime is conditional on the input time.

1. At each depth in the soil, there exists a unique flux pdf describing the area-averaged outflow flux concentration as a function of the cumulative drainage past the outflow boundary, regardless of the rate at which the water is flowing (see (2.5) and (2.63)).

2. The unique flux pdf is stochastic-convective (2.65) within a layer and perfectly correlated at layer boundaries.

3. Field-scale, area-averaged drainage rate can be calculated from a simple water balance model in which drainage flux is a function of the water stored above the point of drainage.

4. Changes in water content within the transport volume are corrected for by shifting the outflow curve along the drainage axis.

These assumptions are discussed in greater detail below.

6.1.1 The Unique Drainage PDF

To avoid having to model transient water flow in the field, Jury (1982) proposed substituting net applied water I or cumulative drainage for time in the travel time pdf. It was hoped this pdf would be approximately independent of the water flow rate, i.e. that solutes would move from the inlet end to the outlet end of the transport volume in a unique manner as a function of the amount of water passing through the system.

Although this model might apply exactly to saturated soil (see Problem 6.3), it cannot be more than an approximate representation of transport through unsaturated soil, because the geometry of the wetted pore space will change at different flow rates. However, it might be reasonably accurate for describing transport through coarse-textured soils that undergo only small changes in water content as the water flux is varied over a large range. In contrast, it might be inaccurate in structured soils, which have large voids that fill only at large flow rates. For such soils, a drainage flux pdf measured at a low flux rate would undoubtedly be less variable than one measured at a high flow rate. This effect has been demonstrated on structured clay soils in undisturbed laboratory columns (Dyson and White, 1986). These authors proposed making the variance of the travel time pdf a function of the irrigation flux rate.

6.1.2 Stochastic-Convective Flow

For many applications in unsaturated soil, the surface area of interest (e.g. an agricultural field) is large, and the depth over which solute transport processes are represented (e.g. a root zone, soil above shallow ground water) is small. Under such conditions, the transverse mixing time will be much longer than the convection time for solutes moving through the surface zone. Consequently, the area-averaged (field scale) travel time pdf may be represented accurately with a stochastic-convective model (2.65) in homogeneous soil, and a with a perfect correlation model (5.25) at the interface between layers in a heterogeneous soil. The stochastic-convective model (2.65) worked extremely well in the top 2 m of soil in a loamy sand field where solute was added over a 0.64 ha surface area (Butters et al., 1990; Butters and Jury, 1990).

6.1.3 Transient Convolution Integral

The model of Jury et al. (1990) will be developed below for macroscopically homogeneous soil in the vertical direction. It can easily be extended to layered soil by making the drainage function and solute flux pdf different in each region. The approach taken in this development will be to begin with the steady-state form of the model (2.5) and to adjust it systematically for transient conditions.

The Steady-State Model

The steady state form of the transfer function equation expressed in terms of net applied water or cumulative drainage $I = J_w t$ is given by (2.5) applied at depth z

$$C^f(z, I) = \int_0^I C^f(0, I - I')\, f^f(z, I')\, dI' \,, \tag{6.1}$$

where the flux pdf is assumed to be stochastic-convective in terms of I. Thus,

$$f^f(z, I) = \frac{\ell}{z} f_\ell^f\left(\ell, \frac{I\ell}{z}\right) \,, \tag{6.2}$$

where ℓ is the reference depth at which the pdf parameters are defined.

Adjustment for Initial Storage

Equation (6.2) is assumed to have been calibrated at a particular steady-state water flux J_w^{ss}, during which time the transport volume had a characteristic average water content θ^{ss}. When the narrow input pulse of solute was added to the surface to measure the flux pdf in the steady-state calibration experiment, a certain volume/area W^{ss} of tracer-free water left the transport volume before the tracer arrived. If W^{ss} is defined relative to the position of the center of mass of the solute, it is approximately equal to

$$W^{ss} = \int_0^z \theta^{ss}(z)dz \approx z\theta^{ss} , \tag{6.3}$$

if θ^{ss} is constant with depth.

If the steady-state experiment is repeated when the initial water storage of the soil profile is significantly different from W^{ss}, a different volume/area W of tracer-free water will precede the appearance of solute at the outlet end. This new antecedent storage volume will be greater or less than W^{ss}, depending on whether the soil profile was wetter or drier than it was in the steady-state experiment. A change in the initial storage will alter the flux pdf (6.2) from the form it had in steady state, even if it is a unique function of I. To a first approximation, we can correct for this change by shifting the steady-state pdf $f^f_{ss}(z,I)$ along the drainage axis by the difference in the amount of initial water storage ΔW between the calibration experiment (now called steady state) and the new experiment,

$$f^f(z,I) = f^f_{ss}(z, I - \Delta W(0)) , \tag{6.4}$$

where the water storage difference at $t = 0$ is equal to

$$\Delta W(0) = \int_0^z \left(\theta(z',0) - \theta_{ss}\right) dz' . \tag{6.5}$$

Figure 6.1 illustrates the correction used in (6.4).

6.1.4 Adjustment for Transient Water Flow

When the cumulative amount of drainage $I(z,t)$ flowing past any depth z in the soil is a nonlinear function of the time, (6.1) must be adjusted in two ways. First, the fundamental pdf (6.2) must be adjusted continually for differences between the profile storage status at the time a given amount of solute is applied to the inlet end, and the steady state storage W^{ss}. Second, the cumulative drainage $I(z,t)$ at depth z will differ from the cumulative water input $I(0,t)$ because of storage changes in the soil. Both effects can be accounted for by changing the independent variable in (6.1) from cumulative water drainage to time. This

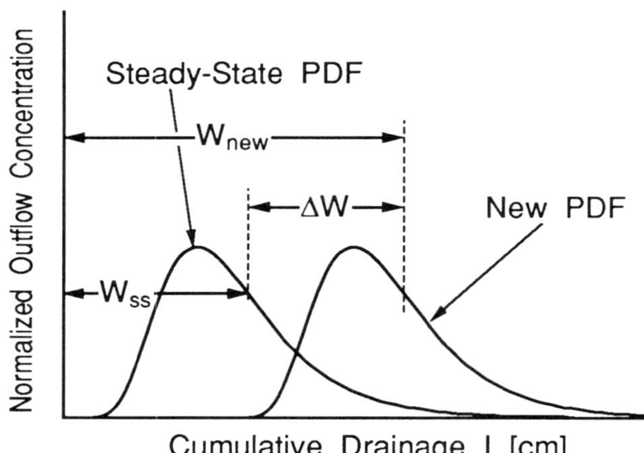

Figure 6.1: Adjustment of the fundamental pdf to compensate for a different amount of initial water storage in the profile.

requires three changes in the integral,

$$\begin{aligned} f^f_{ss}(z, I' - \Delta W(0)) &\longrightarrow f^f_{ss}\Big(z, I(z,t') - \Delta W(I(0,t) - I(0,t'))\Big) \;, \\ dI' &\longrightarrow J_w(z,t')\, dt' \;, \\ C^f(0, I - I') &\longrightarrow C^f\Big(0, I(0,t) - I(0,t')\Big) \;, \end{aligned} \tag{6.6}$$

where $J_w(z,t)$ is the instantaneous drainage rate at depth z and time t, and $\Delta W(I(0,t) - I(0,t'))$ was the water storage difference when the solute which arrived at z at time t and cumulative drainage $I(z,t)$ was added to the surface. By the water continuity equation,

$$I(z,t) = I(0,t) - W(t) + W(0) = I(0,t) - \Delta W(t) \;, \tag{6.7}$$

where

$$\Delta W(t) = \int_0^z \Big(\theta(z',t) - \theta(z',0)\Big)\, dz' \tag{6.8}$$

and

$$I(z,t) = \int_0^t J_w(z,t)\, dt \;. \tag{6.9}$$

All of the water flow parameters are field-scale, area-averaged quantities. Finally, the flux concentration $C^f(z,I)$ on the left side of (6.1) is changed to

$$C^f(z,I) \longrightarrow C^f(z, I(z,t)) \equiv C^f(z,t) \;. \tag{6.10}$$

Thus, the final form of the transient transfer function model is given by

$$C^f(z,t) = \int_0^t C^f(0, I(0,t) - I(0,t')) \\ f^f_{ss}\big(I(z,t') - \Delta W(I(0,t) - I(0,t'))\big)\, J_w(z,t')\, dt' \,. \qquad (6.11)$$

6.2 Water Balance Model

The transfer function solution (6.11) to the field scale transient solute transport problem avoids the complication of modeling local water flux by making the solute pdf a unique function of the field scale drainage. The field scale drainage is described by the one dimensional water conservation law (6.7), and therefore is much easier to predict than local water flux, which may be three dimensional. A simple way to generate the field scale drainage functions $I(z,t)$ and $W(t)$ needed in the model (6.11) is with the layered gravity flow model (Jury et al., 1975). This model assumes that the area-averaged water flux $J_w(z,t)$ at a given depth is a unique function of the water content. To allow the flux to be represented at several depths, the soil is divided into a series of layers of uniform, but time-dependent, water content. The water flow problem is specified by simple mass balance equations for each soil layer

$$\begin{aligned} L_1 \frac{d\theta_1}{dt} &= J_{in}(t) - K_1(\theta_1)\,, \\ L_j \frac{d\theta_j}{dt} &= K_{j-1}(\theta_{j-1}) - K_j(\theta_j)\,, \qquad j = 2, \ldots, n \end{aligned} \qquad (6.12)$$

where $K_j(\theta) := J_w(\theta_j)$ is the drainage rate at the bottom $(z = L_1 + \ldots + L_j)$ of layer j and L_j is the thickness of that layer.

Several investigators have found that the drainage flux at a given depth in the soil can be represented uniquely as a function of the water content above that depth. Black et al. (1969) showed that the measured drainage rate out of a field lysimeter of 4 m^2 surface area and 1.5 m depth filled with *Plainfield* sand could be represented well by the function

$$K(\theta) = K_0 \exp(-\beta(\theta - \theta_0))\,, \qquad (6.13)$$

where $K_0 := K(\theta_0)$ is one point on the curve and β is the slope of $\ln(K)$ versus θ. This model was also used by Nielsen et al. (1973) and Libardi et al. (1980) to represent the drainage rate at a given depth in field plots used to measure unsaturated hydraulic conductivity during redistribution.

Equation (6.12) can be solved rapidly for a large number of layers using (6.13), because this system of equations has an approximate analytic solution for a given time step Δt, if $J_{in}(t)$ is replaced by its average value during each time interval (Jury et al., 1990; also see Problem 6.4). Once the water balance model has been solved over the time period of interest, the transfer function convolution integral (6.11) can be solved numerically for any $C_f(0, I(t))$.

The transient transfer function model was field-tested recently (Jury et al., 1990). In this study, the time-dependent movement of a solute pulse added to a 0.64 ha surface area (Jury et al., 1982) was predicted at five depths (30 to 180 cm) during a winter of erratic rainfall. The transient profiles were calculated with (6.11), using a steady-state measurement of the flux pdf at 30 cm from a later experiment (Butters et al., 1989), and a water balance model generated from air-entry permeameter measurements of $K(\theta)$ (Russo and Bresler, 1980). The agreement between the model (6.11) and the observed solute flux concentrations was quite good, especially considering that the only data taken during the experiment that were used in the simulation were the rainfall rate and the evaporation rate. All other information needed in (6.11) was obtained at a different time from other measurements at the field site (Jury et al., 1990).

6.3 Stochastic Simulation of Transient Water Flow

The simple pesticide pollution screening model described in Chapter 4 assumed that the water input rate to the soil surface was steady. However, pesticide leaching is often the result of transient rainfall events or periodic irrigations, and a season of chaotic water inputs might produce a different residual mass fraction RMF than

The reason that transient effects might be important in pesticide leaching is that the RMF is affected by the amount of time after application that a pesticide spends in the surface soil layer where biodegradation occurs at maximum levels. If, for example, it requires 20 cm to leach a major portion of a mobile chemical below this zone in a particular soil, then a rapidly degrading chemical of this mobility will clearly have a different RMF if 20 cm of rain arrives in two large storms the first week after application, than if the rain arrives uniformly at a much lower flux rate.

Jury and Gruber (1989), in an approach motivated by the study of Small and Mullar (1987), produced a transfer function model of the leaching process under stochastic rainfall input by making the following assumptions:

1. The soil was assumed to have a shallow zone of thickness ℓ within which chemical decayed by a first order reaction (see Section 4.8) with a rate coefficient μ, and below which biodegradation was negligible.

2. The pesticide was everywhere adsorbed to equilibrium with a retardation factor $R = 1 + \rho_b f_{oc} K_{oc}/\theta$.

3. The flux pdf at depth ℓ was assumed to be a unique function of net applied water I to the surface, which was assumed to be equal to the cumulative drainage at all depths (i.e. water storage changes in the transport volume were neglected).

4. The flux pdf was modeled with a gamma distribution. Therefore, for a mobile, inert chemical

$$f_m^f(\ell, I) = \frac{\beta^{1+\alpha} I^\alpha \exp(-\beta I)}{\alpha!} , \tag{6.14}$$

   where the parameters α and β are related by

$$\frac{1+\alpha}{\beta} = L\theta . \tag{6.15}$$

   Thus, the flux pdf of the adsorbing pesticide is, by (4.12)

$$f_a^f(\ell, I) = \frac{1}{\alpha!}\left(\frac{\beta}{R}\right)^{1+\alpha} I^\alpha \exp\left(-\frac{\beta I}{R}\right) . \tag{6.16}$$

5. The cumulative amount of rainfall I reaching the soil surface in time t is modeled as a random variable conditional on t (Eagleson, 1978). The randomness arises from two sources: a random amount of time τ between storms, described by an exponential pdf

$$f_\tau(\tau) = \omega \exp(-\omega\tau) \tag{6.17}$$

   and a random amount of water h per storm, described by a gamma pdf

$$f_h(h) = \frac{\lambda^\kappa h^{\kappa-1} \exp(-\lambda h)}{\Gamma(\kappa)} , \tag{6.18}$$

where $\Gamma(\kappa) := (\kappa - 1)!$ is the Gamma or factorial function (Abramowitz and Stegun, 1970). As shown by Eagleson (1978; see also Problem 6.5), the conditional pdf of I given t may be written as

$$f_P(I \mid t) = \sum_{\nu=1}^{\infty} \frac{(\eta\kappa)^{\nu\kappa} \, I^{(\nu\kappa-1)}}{\Gamma(\nu\kappa)} \exp(-\eta\kappa I) \frac{(\omega t)^{\nu} \exp(-\omega t)}{\nu!} \tag{6.19}$$

where $\eta = \lambda/\kappa$.

The travel time pdf to depth $z = \ell$ for an adsorbing chemical in this system is created by multiplying the probability $f_P(I \mid t)$ that an amount of water will arrive in time t by the probability $f_a^f(\ell, I)$ that the chemical will reach depth ℓ when an amount of water I has been added to the surface. That product gives the probability of reaching depth ℓ after I water has fallen in time t. This must be integrated over all values of I to produce the travel time pdf. Thus

$$f_a^f(\ell, t) = \frac{\int_0^\infty f_P(I \mid t) \, f_a^f(\ell, I) \, dI}{\int_0^\infty \int_0^\infty f_P(I \mid t) \, f_a^f(\ell, I) \, dI dt} , \tag{6.20}$$

where the denominator is required to normalize the pdf to unit area in time (Jury and Gruber, 1989). After insertion of (6.18)–(6.19), the solution to (6.20) is (see Problem 6.7)

$$f_a^f(\ell, t) = \omega \frac{\sum_{\nu=1}^{\infty} \left(\chi^{\nu\kappa} \, \Gamma(\nu\kappa + \alpha) \, (\omega t)^{\nu} \exp(-\omega t)\right) / \left(\Gamma(\nu\kappa) \, \nu!\right)}{\sum_{\nu=1}^{\infty} \left(\chi^{\nu\kappa} \, \Gamma(\nu\kappa + \alpha)\right) / \Gamma(\nu\kappa)} , \tag{6.21}$$

where $\chi := (1 + \beta/\eta\kappa R)^{-1}$ (Jury and Gruber, 1989).

With this new pdf, the RMF which was calculated in Chapter 4 assuming steady state water flux may now be reestimated as a function of the parameters (η, κ, ω) characterizing the uncertainty of the rainfall distribution. For example, the mean and variance of the RMF are calculated from

$$\mathrm{E(RMF)} = \mathrm{E}(\exp(-\mu t)) = \int_0^\infty \exp(-\mu t) \, f_a^f(\ell, t) \, dt , \tag{6.22}$$

$$\mathrm{Var(RMF)} = \mathrm{E}(\exp(-2\mu t)) - \mathrm{E}^2(\exp(-\mu t)) . \tag{6.23}$$

Finally, since the cumulative probability that the RMF will be greater than or equal to a value γ is equal to

$$\begin{aligned} P(\mathrm{RMF} \geq \gamma) &= \mathrm{Prob}\{\exp(-\mu t) \geq \gamma\} \\ &= \mathrm{Prob}\Big\{t \leq t_\gamma = \frac{1}{\mu}\ln\Big(\frac{1}{\gamma}\Big)\Big\} = \int_0^{t_\gamma} f_a^f(\ell, t)\, dt\;, \end{aligned} \tag{6.24}$$

we may integrate (6.21) to produce

$$P(\mathrm{RMF} \geq \gamma) = \frac{\sum_{\nu=1}^{\infty} \chi^{\nu\kappa}\,\Gamma(\nu\kappa + \alpha)\,(1 - \Psi(\nu, \omega t_\gamma))/\Gamma(\nu\kappa)}{\sum_{\nu=1}^{\infty} \chi^{\nu\kappa}\Gamma(\nu\kappa + \alpha)/\Gamma(\nu\kappa)}\;, \tag{6.25}$$

where $\Psi(\nu, \omega t_\gamma)$ is the incomplete gamma function (Abramowitz and Stegun, 1970)

$$\Psi(\nu, \omega t_\gamma) := \int_0^{\omega t_\gamma} \frac{x^\nu\, \exp(-x)}{\nu!}\, dx\;. \tag{6.26}$$

Example 6.1 CLIMATIC VARIABILITY AND PESTICIDE LEACHING

To illustrate the effect of soil and climatic variability on the uncertainty of the estimate of the RMF, we can select a range of values for the pdfs governing the leaching variability of the soil (6.14), the storm frequency (6.17), and the storm height (6.18). Table 6.1 summarizes the range of conditions studied in the simulations below.

***Table 6.1:** Mean and CV of the pdfs representing soil and climatic variability in the four simulations summarized in Table 6.2.*

Model	Soil PDF $f^f(\ell, I)$		Storm Frequency PDF $f_\tau(t)$		Storm Height PDF $f_h(h)$	
	Mean (cm)	CV	Mean (day)	CV	Mean (cm)	CV
A	45	0.6	3.33	1.00	0.85	1.41
B	45	0.4	20.00	1.00	5.10	3.16
C	45	0.6	20.00	1.00	5.10	3.16
D	45	0.4	3.33	1.00	0.85	1.41

In Table 6.1, note that the mean value E(I) of the soil net applied water pdf for mobile chemicals (6.14) is constant (45 cm), as well as the mean water application rate (93 cm/yr). Thus, scenarios A, B, C, and D differ only in how they model the variability of the climate and the soil. Table 6.2 summarizes calculations of (6.24) giving the probability that the RMF will exceed a given value.

Thus, in this calculation, the probability that the RMF would exceed 0.0001 varied between a low of 65.46% under condition D to a high of 87.09% for C. Under a given climatic condition, the

Table 6.2: *Calculated probability (%) of exceeding the specified value of RMF for the compound aldicarb (K_{oc} = 36 $cm^3 g^{-1}$; $\tau_{1/2}$ = 70 d; Jury and Gruber, 1989) leaching below the 1 m surface zone of a fine textured soil ($\theta = 0.45, \rho_b = 1.2, f_{oc} = 0.025$). Each climate has the same annual rainfall of 92.7 cm. (A: variable soil and uniform climate; B: uniform soil and variable climate; C: variable soil and variable climate; D: uniform soil and uniform climate)*

Specified	Climatic and Soil Conditions			
RMF Value	A	B	C	D
0.0001	83.18	66.10	87.09	65.46
0.0010	66.36	48.60	67.05	45.88
0.0100	40.02	27.66	33.77	23.80
0.1000	11.08	8.34	5.00	6.06
0.5000	0.61	0.79	0.06	0.48

soil variability had a far more significant effect than the climatic variability on the variability of the RMF under a given soil condition. This result is consistant with the findings of Jury and Gruber (1989), who observed that climatic variability was an important factor in the uncertainty of the RMF only for those compounds whose mean leaching time to migrate below the surface zone was less than one season. In that case, the distribution of rainfall within a season would affect what fraction of the applied mass moved below the biologically active zone before degrading.

6.4 Scaling Theory

Scaling theory refers to the use of relationships between soil parameters at different locations based on dimensional analysis, geometric arguments, or even ad hoc empirical functions to simplify the problem of describing transport through spatially variable soils (Tillotson and Nielsen, 1984; Sposito and Jury, 1986). The foundation of the method rests on the classic analysis of Miller and Miller (1956), who developed a theory for scaling water flow through unsaturated soils based on the assumption of geometric similitude. Two porous media are said to be geometrically similar if the particles forming the solid matrix of each have shapes and orientations which are identical except in relative magnification (see Fig. 6.2).

A local region of a soil whose solid matrix conforms to the hypothesis of geometric similarity can be characterized uniquely relative to other parts of the soil by a single scaling factor λ defining the local magnification or length scale. Moreover, the hydraulic properties of the local region can be related to the

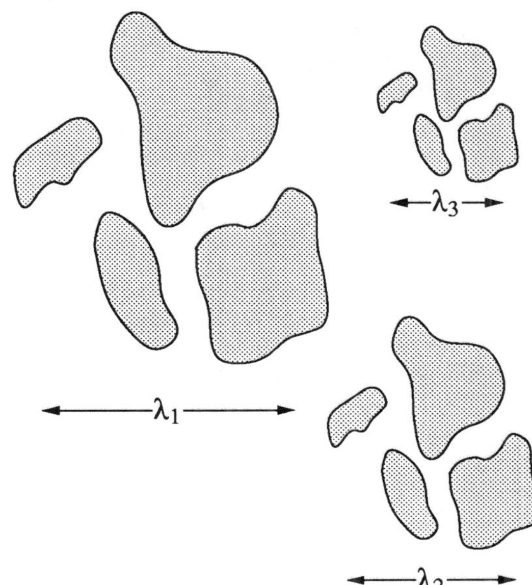

Figure 6.2: Illustration of three regions of a geometrically similar porous medium and the three scaling factors λ defining the local magnification.

hydraulic properties of a reference region within the soil using the relative scale magnification.

As shown by Miller and Miller (1956), all hydraulic and retention properties at a location i where the scale factor is λ_i can be calculated from the values of the properties at the reference location where $\lambda = \lambda^*$. In particular, the unsaturated hydraulic conductivity $K_i(\theta)$ and matric potential $\psi_i(\theta)$ are calculated by

$$K_i(\theta) = \alpha_i^2 K^*(\theta) \ , \tag{6.27}$$

$$\psi_i(\theta) = \alpha_i^{-1} \psi^*(\theta) \ , \tag{6.28}$$

where $\alpha_i = \lambda_i / \lambda^*$ is the magnification of region i relative to the reference region, and $K^*(\theta)$, $\psi^*(\theta)$ are the reference values of the properties. Note that all regions of a geometrically similar soil have the same porosity by definition.

6.4.1 Transport Models Based on Scaling Theory

The potential application of scaling theory to spatially variable soils is clear from (6.27). If a field soil is geometrically similar, then the hydraulic properties of all regions of the soil can be calculated from the hydraulic properties of a single reference region if the scaling factor distribution is known. This idea was first proposed by Peck et al. (1977). Further, the transient water flow or Richards equation (Hillel, 1980) can be scaled and then written in a dimensionless form that depends only on the hydraulic properties of the reference region and the relative scale factor α (Warrick and Amoozegar-Fard, 1979; also see Problem 6.1). This method of modeling soil heterogeneity has been used to develop a scaling theory of solute transport in spatially variable soils, as shown in the next example.

Example 6.2 THE SCALING MODEL OF BRESLER AND DAGAN

In 1979, Dagan and Bresler published two papers (Dagan and Bresler, 1979; Bresler and Dagan, 1979) which used geometric similitude scaling theory to model the spatial variability of the soil hydraulic properties influencing solute transport. In their approach, area-averaged solute transport at the field scale was modeled for steady water flow by treating the soil as a set of locally homogeneous regions i through which water moved by gravity flow according to the equation $J_w = K_i(\theta)$, where $K_i(\theta)$ is the unsaturated hydraulic conductivity of region i. Further, flow of water and solute at the local scale was assumed to be one dimensional, so that the field behaved like a set of parallel soil columns. Thus, when the field was irrigated at a steady rate $J_w = i_0$, each local region reached a different water content θ_i which obeyed

$$\theta_i = \begin{cases} K_i^{-1}(i_0) & , i_0 < K_{i,sat}^* , \\ \theta_{sat} & , i_0 \geq K_{i,sat}^* . \end{cases} \tag{6.29}$$

The field was assumed to be geometrically similar with a lognormal scaling factor distribution. Therefore, the unsaturated hydraulic conductivity in (6.29) obeys (6.27). The reference region was modeled by the functional form

$$K^*(\theta) = K_{sat}^* \left(\frac{\theta}{\theta_{sat}}\right)^{1/\beta} , \tag{6.30}$$

where β is a constant. After combining (6.27), (6.29) and (6.30), one obtains the following relationship between the applied flux i_0 and the water content θ_i at site i

$$i_0 = \alpha_i^2 K_{sat}^* \left(\frac{\theta_i}{\theta_{sat}}\right)^{1/\beta} , \quad \text{if } i_0 < K_{i,sat}^* . \tag{6.31}$$

This equation may be solved for the local water content, with the result

$$\theta_i = \begin{cases} \theta_{sat}\, i_0^{\beta}\, \alpha_i^{-2\beta}\, {K_{sat}^*}^{-\beta} & , i_0 < K_{i,sat}^* , \\ \theta_{sat} & , i_0 \geq K_{i,sat}^* . \end{cases} \tag{6.32}$$

In their first papers, Dagan and Bresler (1979) also assumed that local solute dispersion was negligible (i.e. piston flow was valid in each local soil column). They relaxed this requirement in later applications of the model (Dagan and Bresler, 1983ab; Bresler and Dagan, 1983). If dispersion is neglected, then mobile dissolved chemicals will move at a uniform velocity $V_i = i_0/\theta_i$ in a given region i if there is no ponding, and at $V_i = \alpha_i^2 K_{sat}^* / \theta_{sat}$ if ponding occurs. Therefore, the solute velocity V_i is given by

$$V_i = \begin{cases} {\theta_{sat}}^{-1}\, i_0^{1-\beta}\, \alpha_i^{2\beta}\, {K_{sat}^*}^{\beta} & , \alpha_i > \sqrt{i_0/K_{sat}^*} , \\ {\theta_{sat}}^{-1}\, \alpha_i^2 K_{sat}^* & , \alpha_i \leq \sqrt{i_0/K_{sat}^*} . \end{cases} \tag{6.33}$$

According to (6.33), the local solute velocity V is a function of a set of constant parameters β, i_0, θ_{sat}, K^*_{sat} and a spatially variable parameter α which is described by the lognormal pdf $f(\alpha)$. Thus, the solute travel time flux pdf of the field may be constructed with (4.1) from a local stream tube model which has a single random parameter α

$$f^f(z,t;\alpha) = \delta\Big(t - \frac{z}{V(\alpha)}\Big) \; , \qquad (6.34)$$

if it is assumed that α has no spatial structure. Since α is lognormally distributed by assumption, and the two variables are related by (6.33), we may derive the pdf of V from the pdf of α. Taking the natural logarithm of both sides of (6.33), we obtain

$$\ln(V) = \begin{cases} \gamma + 2\beta Y & , Y > Y_p \; , \\ \eta + 2Y & , Y \leq Y_p \; . \end{cases} \qquad (6.35)$$

where

$$\begin{aligned} Y &= \ln(\alpha) \; , \\ Y_p &= 1/2 \ln(i_0 K^{*\,-1}_{sat}) \; , \\ \gamma &= \ln(\theta_{sat}^{-1}\, i_0^{1-\beta}\, K^{*\,\beta}_{sat}) \; , \\ \eta &= \ln(\theta_{sat}^{-1} K^*_{sat}) \; . \end{aligned} \qquad (6.36)$$

Equation (6.35) may be used to derive the pdf $f_V(V)$ of V from $f_Y(Y)$, since (Grimmett and Welsh, 1986)

$$f_V(V) = f_\alpha(\alpha(V))\frac{d\alpha}{dV} = f_Y(Y(V))\frac{dY}{dV} \; . \qquad (6.37)$$

Thus, if α is lognormally distributed with parameters μ_α, σ_α then

$$f_Y(Y) = \frac{1}{\sqrt{2\pi}\,\sigma_\alpha} \exp\Big(-\frac{(Y-\mu_\alpha)^2}{2\sigma_\alpha^2}\Big) \; , \qquad (6.38)$$

and the pdf of V is given by (see Problem 6.2)

$$f_V(V) = \begin{cases} \dfrac{1}{2\beta V} f_Y\Big(\dfrac{\ln(V)-\gamma}{2\beta}\Big) & , \ln(V) > \gamma + 2\beta Y_p \; , \\ \dfrac{1}{2V} f_Y\Big(\dfrac{\ln(V)-\eta}{2}\Big) & , \ln(V) \leq \gamma + 2\beta Y_p \; . \end{cases} \qquad (6.39)$$

Now that the pdf of V is determined, the flux pdf of the field may be calculated with (4.1) in the form

$$f^f(z,t) = \int_0^\infty \delta\Big(t - \frac{z}{V}\Big)\, f_V(V)\, dV = \frac{z}{t^2}\, f_V\Big(\frac{z}{t}\Big) \; . \qquad (6.40)$$

Therefore, the travel time pdf is, using (6.39)-(6.40)

$$f^f(z,t) = \begin{cases} \dfrac{1}{2\beta t} f_Y\Big(\dfrac{\ln(z/t)-\gamma}{2\beta}\Big) & , \ln\Big(\dfrac{z}{t}\Big) > \gamma + 2\beta Y_p \; , \\ \dfrac{1}{2t} f_Y\Big(\dfrac{\ln(z/t)-\eta}{2}\Big) & , \ln\Big(\dfrac{z}{t}\Big) \leq \gamma + 2\beta Y_p \; . \end{cases} \qquad (6.41)$$

Bresler and Dagan (1979) used the scaling theory model developed above to calculate the mean and variance of the solute concentration for the case of a step function application of solute. They also included the effect of spatial variability in the water application rate i_0.

In later studies, they extended the model to transient water applications and included solute dispersion in the stream tube model for chemical transport (Dagan and Bresler, 1983ab; Bresler and Dagan, 1983).

6.4.2 Tests of Scaling Theory

Despite its elegance, the scaling theory model rests on a shaky theoretical foundation (geometric similitude). Laboratory tests of the scaling relations (6.3) in porous media that approximate the conditions of geometric similitude (for example columns packed with monodisperse glass beads of different diameter) have verified the theory beyond dispute (Miller and Miller, 1956; Klute and Wilkinson, 1958). However, field soil heterogeniety appears to be described poorly by scale magnification. For example, the porosity ϕ of a field soil commonly has a CV of the order of 10% (Jury, 1985). While not large, this variation violates the assumption of geometric similitude, which requires that ϕ be constant.

For this reason, scaling of field hydraulic properties is usually preceded by dividing water content by the local value of ϕ, and scaling the properties as a function of relative saturation $s = \theta/\phi$ (Warrick et al., 1977). This type of scaling, which is no longer geometric similitude, is called *Warrick similitude* in the review of Sposito and Jury (1985).

Tests of Warrick similitude usually consist of measuring several hydraulic properties (e.g. K and ψ) as a function of s at a number of sites around the field, and then comparing the values of α calculated from the comparison of the two properties with their values at one site used as the reference. Warrick et al. (1977) were the first to try this, using the $K(s)$ and $\psi(s)$ data measured at 20 sites over 150 ha in the field study of Nielsen et al. (1973). The two distributions of α, although lognormal in shape and correlated ($r^2 = 0.83$), were quite different. The variance σ_α^2 of $\ln(\alpha)$ for the $f(\alpha)$ distribution obtained from the $\psi(s)$ data was about 0.26, whereas it was 1.36 when calculated from the measurements of $K(s)$ (see Problem 6.9). Similarly, Sharma et al. (1980) found that the scaling factors calculated from two parameters of an infiltration model fitted to local measurements of infiltration rate over a 9.6 ha field were again correlated ($r^2 = 0.83$), but quite different ($\sigma_\alpha^2 = 0.33$ and 1.06).

Russo and Bresler (1980) conducted two different tests of scaling theory, one using $K(s)$ and $\psi(s)$ data, and a second involving infiltration measurements on the same field site. The distributions were much more similar than in previous studies ($\sigma_\alpha^2 = 0.30$ to 0.57) but not as well correlated ($r^2 = 0.57$) except in a narrow range around saturation.

Jury et al. (1987) reanalysed the data from the experiments of Nielsen et al. (1973) and Russo and Bresler (1980), and concluded that no single scaling factor could remove the variability around the field, but that a model involving the use of two local scaling factors removed much of the variability in both studies. Future research may well lead to improvements in the prospects for using some sort of scaling model to model spatial variability.

Problems[2]

Problem 6.1 Apply the geometric similitude scaling relations to the Richards equation

$$\frac{\partial \theta}{\partial t} = \frac{\partial}{\partial z}\left(K(\psi)\frac{\partial \psi}{\partial z}\right) - \frac{\partial}{\partial z}K(\psi) \tag{6.42}$$

describing water flow under transient conditions, to produce an equation in terms of the properties $K^*(\theta)$ and $\psi^*(\theta)$ of the reference location. Define a dimensionless time $T := t/t^*$ and length $X := z/z^*$ and select the parameters t^* and z^* so as to remove the scaling factor α from the reduced equation.

Problem 6.2 Derive the pdf $f_V(V)$ in (6.39) and show that it is normalized.

Problem 6.3 Discuss the conditions under which $f^f(z, I)$ would be a unique function of cumulative water flux in saturated soil for a process that obeyed the CDE.

Problem 6.4 Show that (6.12) has an exact solution over a time period Δt if the input flux is assumed to be a constant over this time interval, and if the drainage function is given by (6.13).

†**Problem 6.5** Derive (6.19), the conditional pdf $f_P(I \mid t)$, and explain why it is not normalized to unit area.

[2] Problems marked with a † are more difficult.

Problem 6.6 Show that the integral form of the steady state transfer function (6.1) modified by (6.4) preserves mass balance.

Problem 6.7 Derive (6.21) from (6.19)–(6.20).

†**Problem 6.8** Derive the conditional pdf $f_P(I \mid t)$ for the special case of one storm per year of height H occurring on the first day of the year. Calculate the travel time pdf corresponding to this pdf.

Problem 6.9 In the paper of Warrick et al. (1977), the authors analyed the data from the field study of Nielsen et al. (1973) using scaling theory. They calculated one set of scaling factors α from the $\psi(\theta)$ data and a second set from the $K(\theta)$ data. The two sets were both lognormally distributed, but had very different parameters (μ_α, σ_α) for the log mean and log variance of α.

Hydraulic Function	μ_α	σ_α
$\psi(\theta)$	-0.13	0.51
$K(\theta)$	-0.62	1.17

Calculate the mode, median, mean value and CV predicted for the saturated hydraulic conductivity K_s (which obeys (6.27)) for both assumed distributions of α. Assume that the reference site has $K^*_{sat} = 10$ cm/hr.

Chapter 7

Stochastic Continuum Modeling

The soil matrix is an exceedingly complicated structure, with typical lengths ranging from less than 10^{-6} m between clay particles, to hundreds of meters for soil types. The complicated structure of the matrix influences many other soil properties like e.g. soil density, water content, energy density of water, and hydraulic conductivity. In every experiment, the soil is explored at a certain scale. Structures with typical lengths much larger than this scale may be considered as constant or as a linear drift, whereas structures with much smaller typical lengths may appear as a random component. In this book we do not deal with the complications that are introduced by variations at scales larger than that of the experiment: we always assume the soil to be homogeneous at scales larger than those that are studied explicitly. The existence of dispersion tells us, however, that the same point of view is not permissible concerning smaller scales, because there would be no dispersion in an absolutely homogeneous system.

Up to now we have characterized the soil as a medium that is homogeneous at the scale of interest and only becomes heterogeneous at a very much smaller scale. It was therefore not necessary to model explicitly the complicated structure of the transport volume, and the complex solute flow paths resulting from this structure were represented with a lumped outflow parameter, such as the dispersivity in the CDE model. In this chapter we will explore different approaches to account for the detailed structure of the transport volume and its contribution to solute movement, at least in a statistical way.

7.1 The N Layer Travel Time PDF

As a first stochastic model of a homogeneous soil we will consider a medium consisting of a large number N of identical homogeneous layers of thickness Δz.

Each of these layers is described by the travel time pdf $f^f(\Delta z, t)$. If we assume that solute entering the inlet end of the transport volume must travel through each layer before reaching the outflow end, then the travel time pdf for the entire transport volume is given by[1]

$$f^f(z,t) = \int_0^\infty \ldots \int_0^\infty \delta\Big(t - \sum_{j=1}^{N} t_j\Big)\, f(t_1,\ldots,t_N)\, dt_1 \ldots dt_N \ , \tag{7.1}$$

where t_j is the travel time through the jth layer and $f(t_1,\ldots,t_N)$ is the joint pdf of $t_1,\ldots,t_N$. The mean and variance of t follow immediately from (5.20)–(5.21) and the assumption that the layers are identical,

$$\mathrm{E}_z(t) = \sum_{j=1}^{N} \mathrm{E}_z(t_j) = N\mu \tag{7.2}$$

$$\mathrm{Var}_z(t) = \sum_{i=1}^{N}\sum_{j=1}^{N} \mathrm{E}_z(t_i t_j) - N^2\mu^2 = \sigma^2 \sum_{i=1}^{N}\sum_{j=1}^{N} \rho_{ij} \tag{7.3}$$

where μ and σ are the mean and standard deviation of the travel time through a single layer of thickness Δz, and ρ_{ij} is the correlation coefficient between the ith and jth layers. The simplest representation of the joint travel time pdf in (7.1) is to assume it is multivariate normal. This case is analysed in the next example.

Example 7.1 MULTIVARIATE NORMALLY DISTRIBUTED TRAVEL TIMES

If we assume that the travel times of individual layers of thickness Δz in a homogeneous soil whose travel time pdf is represented by (7.1) are all normally distributed with mean μ and variance σ^2, and correlated according to

$$\rho_{ij} = \rho^{|i-j|} \ , \tag{7.4}$$

where ρ is a constant, then the joint pdf $f(t_1,\ldots,t_N)$ is said to be multivariate normally distributed (Himmelblau, 1970). In this case, the variance (7.3) is equal to (see Problem 7.1)

$$\frac{\mathrm{Var}_z(t)}{\sigma^2} = N + \frac{2\rho N}{1-\rho} - \frac{2(\rho - \rho^{N+1})}{(1-\rho)^2} \ . \tag{7.5}$$

If we now let $N = z/\Delta z$, and write $\rho^x = \exp(x \ln(\rho))$, (7.5) may be written as

$$\frac{\mathrm{Var}_z(t)}{\sigma^2} = \frac{1+\rho}{1-\rho}\frac{z}{\Delta z} - \frac{2\rho}{(1-\rho)^2}\Big(1 - \exp\Big(-\frac{z}{H_t}\Big)\Big) \ . \tag{7.6}$$

[1]Equation (7.1) may be thought of as a stochastic stream tube model, with the δ-function inside the integral representing the local stream tube flux pdf.

The parameter

$$H_t = -\frac{\Delta z}{\ln(\rho)} \tag{7.7}$$

may be thought of as a characteristic length scale for the travel time, representing the distance over which the transport process changes from stochastic-convective to convective-dispersive. Moreover, the macrodispersivity λ, defined by

$$\lambda = \frac{D}{V} = \frac{V^2}{2z}\,\mathrm{Var}_z(t) = \frac{\Delta z^2}{2\mu^2 z}\,\mathrm{Var}_z(t)\ , \tag{7.8}$$

may be combined with (7.6), to produce

$$\lambda = \frac{(1+\rho)}{(1-\rho)}\gamma - \frac{2\gamma\Delta z\rho}{z(1-\rho)^2}\left(1 - \exp\left(-\frac{z}{H_t}\right)\right)\ , \tag{7.9}$$

where

$$\gamma = \frac{1}{2}\left(\frac{\sigma}{\mu}\right)^2 \Delta z\ . \tag{7.10}$$

As shown in Problem 7.2, γ is equal to the dispersivity at $z = \Delta z$. The dispersivity (7.9) is plotted as a function of z for various values of ρ in Figure 7.1.

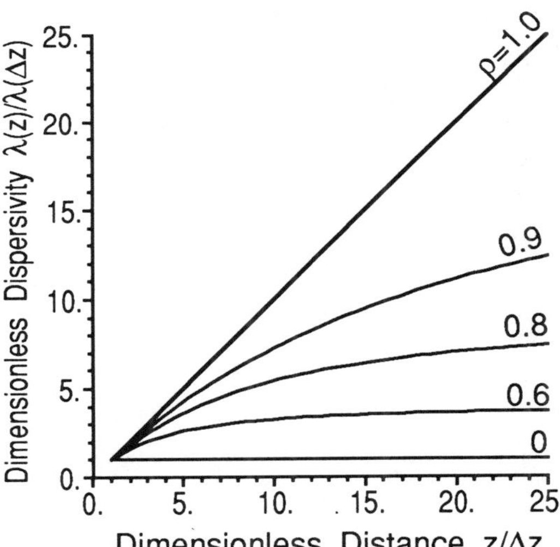

Figure 7.1: Dispersivity growth with distance in a homogeneous medium with a correlation coefficient ρ between layers of thickness Δz.

The previous example illustrates the relationship between travel time correlations and macrodispersion, and is a useful conceptual aid to understanding the relation between the continuum approaches of previous chapters and the stochastic continuum model. However, the correlated travel time model above has several serious limitations that prevent it from being a practical tool for predicting solute transport over great distances. First, the normal distribution is unsuitable for representing travel time pdfs, both because real travel time pdfs tend to be skewed (Jury, 1985), and because for large CV, a normally distributed travel time pdf includes negative travel times. Second, the correlation coefficient ρ is not measurable, and can only be estimated by fitting the expression (7.6) to the variance at several depths.

If a porous medium was macroscopically homogeneous along the direction of flow, then all layers would have identical travel time pdfs. A direct measurement of the pdf of a surface layer would therefore define all the pdf parameters, leaving only ρ to be fitted from subsequent observations. In practice, however, field data is likely to be sufficiently uncertain that the subsequent observation would have to be at a depth considerably below the surface layer (e.g. deeper than 5 m) to allow the parameter to be estimated accurately (Jury and Sposito, 1985). Since this distance is likely to be comparable to the entire transport volume for many problems of interest in unsaturated soil, a correlated travel time model would be superfluous, since its calibration would require a direct measurement of $f^f(z,t)$.

For this reason, a more promising approach to stochastic continuum modeling would seem to be to use a process model to relate the travel time (or travel depth) to some independently measurable property of the porous medium. This philosophy has guided the major research efforts in stochastic continuum modeling to date. Before describing the procedures used in this approach, it is worthwhile to review some concepts relevant to random functions and stochastic processes.

Excursus Random Functions and Stochastic Processes

A *random variable* ω is defined by the set of values D_ω that it can take and by its probability distribution $f(\omega)$ on D_ω (Van Kampen, 1976).

A *random function* is a function that depends on non-random as well as on random variables. It is called a *stochastic process* if one of the non-random variables is time.[2] For simplicity we will only consider random functions $Z(x;\omega)$ that depend on one non-random, real variable x and one random, real variable ω. This random function may be regarded as an *ensemble* of real functions $Z(x)$ which are "numbered" by the random variable ω (Van Kampen, 1976, 1981; Papoulis, 1984), i.e. for a specific ω_0, $Z(x;\omega_0)$ is a real function and is called a *representation* of the random function. For a specific x_0, $Z(x_0;\omega)$ is a real random variable and $Z(x_0;\omega_0)$ finally is a real number.

The *expectation* or *ensemble average* of the random function $Z(x;\omega)$ is defined as the

[2] Stochastic processes involving one time variable and at least one space variable are often referred to as *random fields* (Van Kampen, 1981).

average over the random variable ω,

$$\overline{Z}_1(x) := \langle \boldsymbol{Z}(x;\omega)\rangle := \int_{D_\omega} \boldsymbol{Z}(x;\omega) f(\omega)\, d\omega \ , \tag{7.11}$$

and its *moments* of order n are defined as

$$\begin{aligned} \overline{Z}_n(x_1, x_2, \ldots, x_n) &:= \langle \boldsymbol{Z}(x_1;\omega)\boldsymbol{Z}(x_2;\omega)\cdots \boldsymbol{Z}(x_n;\omega)\rangle \\ &:= \int_{D_\omega} \boldsymbol{Z}(x_1;\omega)\boldsymbol{Z}(x_2;\omega)\cdots \boldsymbol{Z}(x_n;\omega) f(\omega)\, d\omega \ . \end{aligned} \tag{7.12}$$

(The moments (2.28) of a random variable may be interpreted as a special case of (7.12) where $\boldsymbol{Z}(x;\omega)$ does not depend on x.)

A random function is called *stationary* (or strictly stationary) if its moments only depend on the differences $x_2 - x_1, \ldots, x_n - x_1$, i.e. if the moments of $\boldsymbol{Z}(x;\omega)$ and $\boldsymbol{Z}(x - x_0;\omega)$ are identical for any fixed x_0. The moments of a stationary function are much easier to determine, because it is usually sufficient to measure $\boldsymbol{Z}(x)$ on an interval $[x_0 - \Delta x, x_0 + \Delta x]$ instead of the whole x-axis. One therefore often assumes that the random function under consideration is stationary.

To determine a random function completely, its moments of all orders must be known. This is clearly not attainable in practice, because it would involve an infinite number of measurements even for a stationary process. In applications, one considers therefore only the first few moments, often only the first and the second, and introduces the notation of a *weakly stationary* random function whose first moment, the expectation $\overline{Z}_1$, is independent of x and whose second moment $\overline{Z}_2$ depends only on relative values $\xi := x_2 - x_1$, called *lag distances.* The *covariance*[3] or auto-covariance $\sigma_Z(\xi)$ of a weakly stationary random function $\boldsymbol{Z}(x)$[4] is defined as

$$\begin{aligned} \sigma_Z(\xi) = \mathrm{Cov}_Z(\xi) &:= \Big\langle \big(\boldsymbol{Z}(x) - \overline{Z}_1\big)\big(\boldsymbol{Z}(x+\xi) - \overline{Z}_1\big)\Big\rangle \\ &= \langle \boldsymbol{Z}(x)\boldsymbol{Z}(x+\xi)\rangle - \overline{Z}_1^2 = \overline{Z}_2(\xi) - \overline{Z}_1^2 \ , \end{aligned} \tag{7.13}$$

and the *variance* as

$$\mathrm{Var}(\boldsymbol{Z}) := \sigma_Z(0) \ . \tag{7.14}$$

In the geostatistical approach (Journel and Huijbregts, 1978), one often uses the *semi-variogram*

$$\gamma(\xi) := \frac{1}{2}\Big(\mathrm{Var}\big(\boldsymbol{Z}(x+\xi) - \boldsymbol{Z}(\xi)\big)\Big) = \mathrm{Cov}(0) - \mathrm{Cov}(\xi) \tag{7.15}$$

instead of the covariance. Both functions, covariance and semi-variogram, are equivalent descriptions of the spatial structure of weakly stationary random functions. However, if they are

[3]Sometimes $\sigma_Z(\xi)$ is referred to as correlation or auto-correlation (Van Kampen, 1976). We prefer, however, to reserve this expression for the normalized form (7.16).

[4]In the following, we will suppress the variable ω when it is not needed explicitly to make the notation more compact. The random function $\boldsymbol{Z}(x)$ and its realizations $Z(x)$ will still be distinguishable, however, as the random function is always denoted by boldface symbols.

estimated from actual data, the fitted models frequently do not yield the theoretical values $\sigma_Z(0) = Var(\boldsymbol{Z})$ and $\gamma(0) = 0$ at $\xi = 0$. The origin of this discrepancy is at least partly due to the finite distance between measurements which makes the detection of small scale structures impossible. They contribute to the estimates of $\sigma_Z(0)$ and $\gamma(0)$, but not to those for lags greater than 0. The covariance and the semi-variogram for an idealized weakly stationary random function are shown in Figure 7.2.

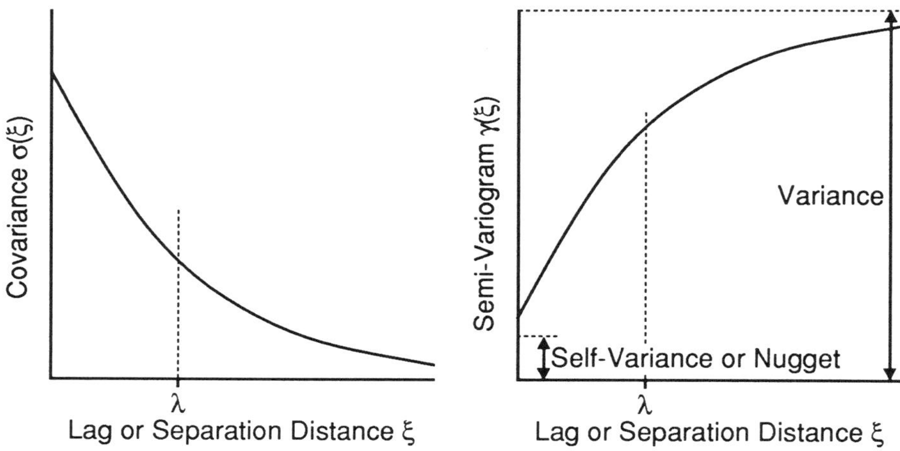

Figure 7.2: *The covariance function and semi-variogram of an idealized weakly stationary random function. The integral scale λ, also called range, corresponds to the spatial distance over which the random function is correlated.*

The *correlation* of the random function $\boldsymbol{Z}(x)$ is obtained by normalizing the covariance with the variance

$$\rho_Z(\xi) := \frac{\sigma_Z(\xi)}{\sigma_Z(0)} = \frac{\mathrm{Cov}_Z(\xi)}{\mathrm{Var}(\boldsymbol{Z})} \ . \tag{7.16}$$

One may define different correlation lengths as a measure for the persistence of structures in $\boldsymbol{Z}(x)$. We will later use the *integral correlation length* or *integral scale* defined as

$$\lambda_Z := \int_0^\infty \rho_Z(\xi)\, d\xi \ . \tag{7.17}$$

(If the correlation structure of $\boldsymbol{Z}(x)$ is very persistent, e.g. if $\rho_Z(\xi) \propto 1/\xi$, the integral (7.17) and consequentely the integral correlation length may not exist.)

The expectation operator $\langle . \rangle$ in (7.11) is defined as an average over the random variable ω, i.e. over all realizations of $\boldsymbol{Z}(x;\omega)$. In many applications only one, or at most a few realizations of a random function are actually available, and it is in general not possible to calculate the moments by (7.12). It is necessary in such cases to restrict the analysis to a subclass of random functions which are called *ergodic*, for which the ensemble average is equal to an appropriate average of the realization $Z(x)$ over the deterministic variable x. For a stationary random function with a finite integral correlation length, this average may be defined by

$$\langle \boldsymbol{Z}(x) \rangle = \lim_{\Delta x \to \infty} \frac{1}{2\Delta x} \int_{-\Delta x}^{\Delta x} Z(x - x')\, dx' \ . \tag{7.18}$$

Ergodicity, like stationarity, is a property that is usually assumed for a random function without experimental justification.

The distinction between a random function and a stochastic process is that the latter is a function of the evolving time variable. Therefore, such notions as casuality and transitions from one state to the other appear as part of the description of a stochastic process. In analogy to random variables we define for the stochastic process $\boldsymbol{Z}(t)$ the *probability distribution function* of order n (Papoulis, 1984)

$$F_Z(z_1, t_1; \ldots; z_n, t_n) := \mathrm{Prob}\Big\{ \bigcap_{j=1}^{n} (\boldsymbol{Z}(t_j) \le z_j) \Big\} \ , \tag{7.19}$$

and the *probability density function* (pdf) of order n

$$f_Z(z_1, t_1; \ldots; z_n, t_n) := \frac{\partial^n F_Z(z_1, t_1; \ldots; z_n, t_n)}{\partial z_1 \cdots \partial z_n} \ , \tag{7.20}$$

where $\bigcap_{j=1}^{n} e_j$ denotes the intersection of the events e_j. The *conditional pdf* of the stochastic process $\boldsymbol{Z}(t)$ is

$$f_Z(z_n, t_n \mid z_1, t_1; \ldots; z_{n-1}, t_{n-1}) := \frac{f_Z(z_1, t_1; \ldots; z_n, t_n)}{f_Z(z_1, t_1; \ldots; z_{n-1}, t_{n-1})} \ , \tag{7.21}$$

where by convention $t_1 < \ldots < t_n$. The conditional pdf (7.21) is also called *transition probability* because it may be interpreted as the pdf of the transition of a system to the state (z_n, t_n) if it is known to have been in the states $(z_1, t_1) \ldots (z_{n-1}, t_{n-1})$. (The transition probability is actually a density function. Outside mathematics it is common practice, however, to refer to "probability densities" as "probabilities".)

The stochastic process $\boldsymbol{Z}(t)$ is called a *Markov process* if (Van Kampen, 1981)

$$f_Z(z_n, t_n \mid z_1, t_1; \ldots; z_{n-1}, t_{n-1}) = f_Z(z_n, t_n \mid z_{n-1}, t_{n-1}) \ . \tag{7.22}$$

It follows from (7.21)–(7.22) that the pdf $f_Z(z_1, t_1)$ of a Markov process is fully determined by its pdf $f_Z(z_0, t_0)$ at time $t_0 < t_1$ and by the transition probability $f_Z(z_1, t_1 \mid z_0, t_0)$

$$f_Z(z_1, t_1) = \int_{-\infty}^{\infty} f_Z(z_0, t_0) f_Z(z_1, t_1 \mid z_0, t_0)\, dz_0 \ . \tag{7.23}$$

No knowledge of the process is required for times $t < t_0$.

7.2 Travel Time Representation in Terms of Solute Velocity

Simmons (1986b) developed a continuum representation of solute macrodispersion by writing the random travel time from $z = 0$ to z as

$$t(z) = \int_0^z \frac{dz'}{v(z')} \,, \tag{7.24}$$

where $v(z)$ is the velocity of the solute particle with travel time $t(z)$. We assume that $1/v(z)$ is a random function, and decompose it into the sum

$$\frac{1}{v(z)} = \frac{1}{V} + \frac{1}{\zeta(z)} \,, \tag{7.25}$$

where V is the harmonic mean value of $v(z)$, and $1/\zeta(z) := 1/v(z) - 1/V$ is the zero-mean, random fluctuation of $1/v(z)$ about $1/V$. The mean value of $t(z)$ is thus

$$\mathrm{E}_z(t) = \int_0^z \frac{dz'}{V} = \frac{z}{V} \,. \tag{7.26}$$

Similarly, if we assume that $1/v(z)$ is a weakly stationary random function, the variance of t is

$$\mathrm{Var}_z(t) = \int_0^z \int_0^z \sigma^2 \rho\Big(\frac{1}{\zeta(z')\zeta(z'+h)}\Big)\, dz' dh \,, \tag{7.27}$$

where σ^2 is the variance of $1/\zeta(z)$.

Thus, with (7.27), the task of characterizing the z dependence of the travel time variance has been reduced to characterizing the autocovariance function of the fluctuation of the inverse velocity. Using (7.27), we may define a generalized macrodispersivity $\lambda(z)$ as (Simmons, 1986b)

$$\lambda(z) = \frac{V^2}{2} \frac{d\,\mathrm{Var}_z(t)}{dz} = \frac{\mathrm{CV}^2}{v} \int_0^z \rho\Big(\frac{1}{\zeta(z)\zeta(z+h)}\Big)\, dh \,, \tag{7.28}$$

where CV is the coefficient of variation of $1/v$. To allow (7.28) to be used in a practical manner, one must then model the solute velocity in terms of parameters whose spatial autocovariance can be measured.

A disadvantage of this approach is that the velocity $v(z)$ in (7.24) is one-dimensional, whereas local solute velocity fluctuations may occur in any direction relative to the mean flow. Therefore, if the velocity field has a correlation structure, the covariance of the velocity fluctuation in (7.28) may not be independently measurable. An alternative approach is to develop a description of solute transport in terms of the random trajectory of a solute particle moving through a medium with a random velocity field. This approach is developed in the next section.

7.3 Homogeneous Stochastic Medium

In this section we will consider soil as a globally homogeneous, m-dimensional stochastic medium. Based on the experimental design, we will identify two length scales: the *macroscale* Λ_f above which the medium is homogeneous and the *microscale* Λ_d below which the medium is averaged by the measuring process and no detailed description is available. We will refer to the scale between the micro- and the macroscale as the *mesoscale*. We assume that the velocity field $\boldsymbol{U}(\boldsymbol{x})$ is a stochastic process, stationary in space and time, that does not depend on the solute concentration. Solute transport through this soil is a linear process and it is therefore sufficient to describe the movement of a single tracer pulse.

Excursus ARBITRARY INPUT FUNCTION

If the solute is not applied or present as a single pulse, the resulting concentrations may be written as a superposition of pulse responses that are properly shifted and scaled (see (2.4)).

Arbitrary Input Flux Concentration

Let $f^f(\boldsymbol{x},t)$ be the solute flux concentration that is caused by the input flux concentration $\delta(\boldsymbol{x},t)$, i.e. by the application of a unit flux pulse at $\boldsymbol{x} = 0$ at time $t = 0$. The solute flux concentration $C^f(\boldsymbol{x},t)$ due to an arbitrary flux input $g^f(\boldsymbol{x},t)$ may then be written as

$$C^f(\boldsymbol{x},t) = \int_0^t \int_{-\infty}^{\infty} \int_{-\infty}^{\infty} \int_{-\infty}^{\infty} g^f(\boldsymbol{x}',t') f^f(\boldsymbol{x}-\boldsymbol{x}',t-t')\, dx'dy'dz'dt' \ , \qquad (7.29)$$

where $\boldsymbol{x} = (x,y,z)$ is a point in space. In a given situation the input concentration $g^f(\boldsymbol{x},t)$ may not depend on all space coordinates and time which would simplify the integration in (7.29) considerably.

Arbitrary Initial Resident Concentration (Initial Value Problem)

Let $f^r(\boldsymbol{x},t)$ be the solute resident concentration that is caused by the initially present resident concentration $\delta(\boldsymbol{x},t)$, i.e. by the presence of a unit resident pulse at $\boldsymbol{x} = 0$ at time $t = 0$. The solute resident concentration $C^r(\boldsymbol{x},t)$ due to an arbitrary initial resident concentration $g^r(\boldsymbol{x})$ may then be written as

$$C^r(\boldsymbol{x},t) = \int_{-\infty}^{\infty} \int_{-\infty}^{\infty} \int_{-\infty}^{\infty} g^r(\boldsymbol{x}') f^f(\boldsymbol{x}-\boldsymbol{x}',t)\, dx'dy'dz' \ . \qquad (7.30)$$

Formally, we could also write (7.30) for a time dependent resident input $g^r(\boldsymbol{x}, t)$, however, it would be difficult to imagine an experimental situation where this description would apply.

We will split the velocity into two components that correspond to length scales $\Lambda_f < \lambda$ (macroscale) and $\Lambda_d < \lambda < \Lambda_f$ (mesoscale)

$$\boldsymbol{U}(\boldsymbol{x}) = \overline{\boldsymbol{u}} + \boldsymbol{u}_f(\boldsymbol{x}) \,, \tag{7.31}$$

where $\overline{\boldsymbol{u}}$ is the mean velocity and $\boldsymbol{u}_f$ the zero mean velocity fluctuation.[5] The fluctuation $\boldsymbol{u}_f$ is assumed to be a Gaussian process with covariance tensor

$$\boldsymbol{\sigma}_{u_f}(\boldsymbol{x}, \boldsymbol{y}) := \langle \boldsymbol{u}_f(\boldsymbol{x})\, \boldsymbol{u}_f(\boldsymbol{y}) \rangle \,. \tag{7.32}$$

The trajectory $X(t)$[6] of a particle that starts at $(\boldsymbol{x} = 0, t = 0)$ is a realization of the stochastic process $\boldsymbol{X}(t)$ which is defined by

$$\boldsymbol{X}(t) = \int_0^t \boldsymbol{U}(\boldsymbol{X}(t'))\, dt' + \boldsymbol{X}_d(t) \,, \tag{7.33}$$

where $\boldsymbol{X}_d(t)$ is a Gaussian process with zero mean and covariance $2\boldsymbol{d}_d t$ that summarizes the influence of fluctuations on scales $\lambda < \Lambda_d$ (microscale). Using (7.31), we may split (7.33) into three terms which correspond to the three length scales

$$\begin{aligned} \boldsymbol{X}(t) &= \langle \boldsymbol{X}(t) \rangle + \boldsymbol{X}_f(t) + \boldsymbol{X}_d(t) \\ &= \overline{\boldsymbol{u}}t + \int_0^t \boldsymbol{u}_f(\boldsymbol{X}(t'))\, dt' + \boldsymbol{X}_d(t) \,. \end{aligned} \tag{7.34}$$

This integral equation is very difficult to solve because $\boldsymbol{u}_f$ depends on the trajectory $\boldsymbol{X}(t)$. There are different approaches for obtaining approximate solutions of (7.34). In numerical studies, realizations of the velocity field $\boldsymbol{U}(\boldsymbol{x})$ are generated and trajectories are obtained by tracking particles as they move in this field (Tompson et al., 1988). Analytical methods, on the other hand, replace $\boldsymbol{X}(t)$ in the argument of $\boldsymbol{u}_f$ by approximations of different orders (Phythian, 1975; Dagan, 1984). We will follow Dagan (1984) and use a first order approximation by replacing $\boldsymbol{X}(t)$ by its expectation $\langle \boldsymbol{X}(t) \rangle = \overline{\boldsymbol{u}}t$, which simplifies (7.34) to

$$\boldsymbol{X}(t) = \overline{\boldsymbol{u}}t + \int_0^t \boldsymbol{u}_f(\overline{\boldsymbol{u}}t')\, dt' + \boldsymbol{X}_d(t) \,. \tag{7.35}$$

[5] Processes at scales $\lambda < \Lambda_d$ are not represented as velocities, but will be summarized later as a diffusive component of the particle trajectories.

[6] The realization $X(t) = (x(t), y(t), z(t))$ is a vector valued function. However, we denote it with plain symbols to distinguish it from the stochastic process $\boldsymbol{X}(t)$.

The trajectory $\boldsymbol{X}(t)$ is a Gaussian process because $\boldsymbol{X}_d$ and $\boldsymbol{u}_f$ are Gaussian by assumption and the integral is a linear operator on $\boldsymbol{u}_f$. Because $\boldsymbol{X}_f$ and $\boldsymbol{X}_d$ are uncorrelated, the covariance tensor of $\boldsymbol{X}(t)$ may be written as[7]

$$\begin{aligned}\sigma_{X,jk} &:= \langle \boldsymbol{X}_{f,j}(t)\boldsymbol{X}_{f,k}(t)\rangle + 2d_{d,jk}t \\ &= \Big\langle \int_0^t \int_0^t \boldsymbol{u}_f(\overline{\boldsymbol{u}}t_1)\,\boldsymbol{u}_f(\overline{\boldsymbol{u}}t_2)\,dt_1 dt_2 \Big\rangle + 2d_{d,jk}t \qquad (7.36)\\ &= \int_0^t \int_0^t \sigma_{u_f,jk}(\overline{\boldsymbol{u}}t_1, \overline{\boldsymbol{u}}t_2)\,dt_1 dt_2 + 2d_{d,jk}t \ ,\end{aligned}$$

where the definition (7.32) was used. We have now calculated the covariance of the process $\boldsymbol{X}(t)$ in terms of the covariance of the velocity fluctuations.

We may now write the pdf of the trajectory $\boldsymbol{X}(t)$ as (Dagan, 1987)

$$f(X(t)) = \frac{1}{\sqrt{(2\pi)^m |\boldsymbol{\sigma}_X|}} \exp\Big(-\frac{1}{2}\sum_{j=1}^m \sum_{k=1}^m (X_j(t) - \overline{u}_j t)\sigma^{-1}_{X,jk}(X_k(t) - \overline{u}_k t)\Big) \ , \tag{7.37}$$

where $m = 2, 3$ is the dimension at which the transport of solutes is considered. The pdf (7.37) of the trajectories may also be interpreted as a pdf of the particle positions as a function of space and time

$$f(\boldsymbol{x}, t) = \frac{1}{\sqrt{(2\pi)^m |\boldsymbol{\sigma}_X|}} \exp\Big(-\frac{1}{2}\sum_{j=1}^m \sum_{k=1}^m (x_j - \overline{u}_j t)\sigma^{-1}_{X,jk}(x_k - \overline{u}_k t)\Big) \ . \tag{7.38}$$

One can show by direct substitution (see Problem 7.4) that (7.38) satisfies the generalized convection-dispersion equation

$$\frac{\partial f}{\partial t} + \sum_{j=1}^m \Big(V_j \frac{\partial f}{\partial x_j} - \sum_{k=1}^m D_{jk} \frac{\partial^2 f}{\partial x_j \partial x_k}\Big) = 0 \ , \tag{7.39}$$

where the convection velocity and the macrodispersion tensor are defined as

$$V_j := \frac{d\langle \boldsymbol{X}_j(t)\rangle}{dt} = \overline{u}_j \ , \tag{7.40}$$

$$D_{jk} := \frac{1}{2}\frac{d\sigma_{X,jk}}{dt} \ . \tag{7.41}$$

[7]The comma in the subscript does not stand for a derivative as it is customary in tensor notation, but only divides the indices denoting the variable and the space directions.

In general (7.39) is not a CDE as defined in Chapter 2 because the macrodispersion tensor (7.41) is usually not constant.

The aim of stochastic continuum theory is the calculation of the macroscopic transport parameters $\boldsymbol{V}$ and $\boldsymbol{D}$ from observable quantities. For solute transport through saturated media, e.g. in groundwater, one usually measures the hydraulic conductivity tensor $\boldsymbol{K}(\boldsymbol{x})$ and calculates the transport parameters from the mean and covariance tensor of $\boldsymbol{Y} := \ln(\boldsymbol{K}/K_0)$, where K_0 is a characteristic value of $|\boldsymbol{K}(\boldsymbol{x})|$ (Dagan, 1982, 1984, 1987; Gelhar and Axness, 1983). Dagan (1984) obtained analytic expressions for the dispersion tensor $\boldsymbol{D}$ in two and three dimensions for an isotropic exponential covariance function of $\boldsymbol{Y}$

$$\sigma_Y(\boldsymbol{x}) = \sigma_Y^2 \exp\Big(-\frac{|\boldsymbol{x}|}{\lambda_Y}\Big) , \tag{7.42}$$

where σ_Y^2 is the variance (7.14) and λ_Y the integral scale (7.17) of $\boldsymbol{Y}$.

We are not going to reproduce the calculations of Dagan (1984) but simply cite the result for three dimensional transport. He calculated the covariance tensor (7.36) of the particle positions as

$$\begin{aligned}
\sigma_{X,xx} &= 2\sigma_Y^2\lambda_Y^2\Big[\tau - \frac{8}{3} + \frac{4}{\tau} - \frac{8}{\tau^3} + \frac{8}{\tau^2}\Big(1 + \frac{1}{\tau}\Big)\exp(-\tau)\Big] + 2d_L t , \\
\sigma_{X,yy} &= 2\sigma_Y^2\lambda_Y^2\Big[\frac{1}{3} - \frac{1}{\tau} + \frac{4}{\tau^3} - \Big(\frac{4}{\tau^3} + \frac{4}{\tau^2} + \frac{1}{\tau}\Big)\exp(-\tau)\Big] + 2d_T t , \\
\sigma_{X,zz} &= \sigma_{X,yy} , \\
\sigma_{X,jk} &= 0 \quad , \quad j \neq k .
\end{aligned} \tag{7.43}$$

where $\tau := \overline{u}t/\lambda_Y$, and d_L and d_T are the local longitudinal and transverse dispersion coefficients, respectively, and where the coordinate system is chosen so that the mean flow is along the x-direction ($\overline{\boldsymbol{u}} = (\overline{u}, 0, 0)$). From this we may calculate the macrodispersion tensor D_{jk} by (7.41). We obtain

$$\begin{aligned}
D_{xx} &= \sigma_Y^2\lambda_Y\overline{u}\Big[1 - \frac{4}{\tau^2} + \frac{24}{\tau^4} - \Big(\frac{24}{\tau^4} + \frac{24}{\tau^3} + \frac{8}{\tau^2}\Big)\exp(-\tau)\Big] + d_L , \\
D_{yy} &= \sigma_Y^2\lambda_Y\overline{u}\Big[\frac{1}{\tau^2} - \frac{12}{\tau^4} + \Big(\frac{12}{\tau^4} + \frac{12}{\tau^3} + \frac{5}{\tau^2} + \frac{1}{\tau}\Big)\exp(-\tau)\Big] + d_T ,
\end{aligned} \tag{7.44}$$

for the longitudinal and the transverse components, respectively. The off-diagonal elements—D_{jk}, $j \neq k$—are zero.

Figures 7.3 and 7.4 show plots of the time dependence of the longitudinal and transverse elements of the covariance tensor (7.43) and of the macrodispersion tensor (7.41) for a stochastic medium that is similar to the Borden aquifer (Freyberg, 1986).

Based on the dimensionless time $\tau = \overline{u}t/\lambda_Y$—the number of integral correlation lengths λ_Y that the center of mass has traveled—we will consider three approximations to (7.43) and (7.44). These are (i) $t \gg \lambda_Y/\overline{u}$ when the pulse has already moved through many correlation lengths, (ii) $t < \lambda_Y/\overline{u}$ when the pulse has traveled less than a correlation length, and (iii) $t \ll \lambda_Y/\overline{u}$ (see Figures 7.3 and 7.4).[8]

For large dimensionless times τ, we neglect all terms in (7.43)–(7.44) that contain powers of τ^{-1} and obtain for the covariance tensor

$$\tau \gg 1: \qquad \sigma_{X,xx} \approx 2(d_L + \sigma_Y^2 \lambda_Y \overline{u})t - \frac{16}{3}\sigma_Y^2 \lambda_Y^2, \quad \sigma_{X,yy} \approx 2d_T t + \frac{2}{3}\sigma_Y^2 \lambda_Y^2 \qquad (7.45)$$

and for the macrodispersion tensor

$$\tau \gg 1: \qquad D_{xx} \approx d_L + \sigma_Y^2 \lambda_Y \overline{u}, \quad D_{yy} \approx d_T \ . \qquad (7.46)$$

Because the macrodispersion tensor approaches a constant value in this limit, the spreading of a solute pulse in an isotropic stochastic continuum may be described by a convection-dispersion process. As (7.46) shows, the macrodispersion tensor (7.41) of this process is usually highly anisotropic because the longitudinal component D_{xx} depends on large-scale parameters while the transverse components $D_{yy} = D_{zz}$ only depend on small-scale fluctuations. This is true even in a locally isotropic medium with $d_L = d_T$. Figures 7.3–7.4 further show that the solute plume must travel many integral correlation lengths before the macrodispersion tensor becomes constant. In the case where the hydraulic conductivity is not isotropic, this may well prevent transport from becoming a convection-dispersion process before the solute plume reaches the boundary of the aquifer (Matheron and De Marsily, 1980).

For small dimensionless times τ, the covariance tensor may be approximated by the first two terms of the Taylor series of (7.43) at $\tau = 0$,

$$\tau < 1: \qquad \sigma_{X,xx} \approx 2d_L t + \frac{8}{15}\sigma_Y^2 \overline{u}^2 t^2, \quad \sigma_{X,yy} \approx 2d_T t + \frac{1}{15}\sigma_Y^2 \overline{u}^2 t^2 \ . \qquad (7.47)$$

[8]The validity of the conclusions that we are drawing for these approximations are naturally limited by the validity of (7.43).

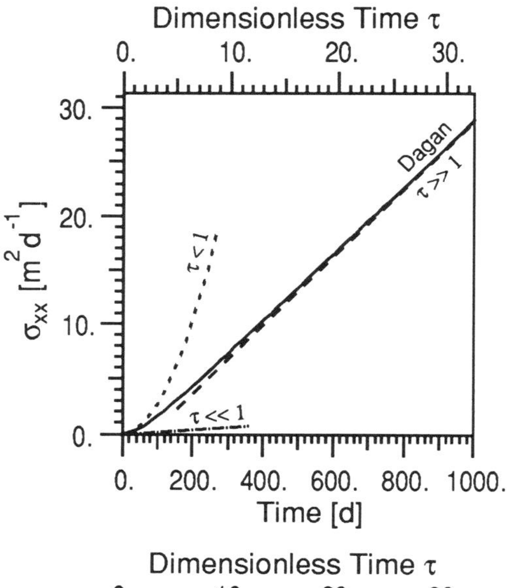

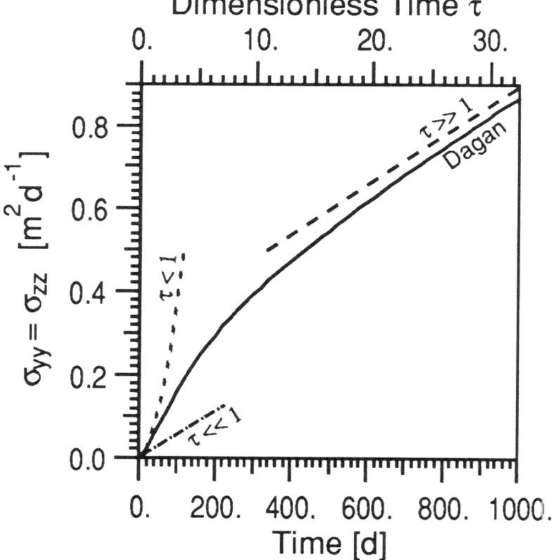

Figure 7.3: *Elements of the covariance tensor (7.43) in an aquifer with an isotropic, exponentially decaying covariance of the log hydraulic conductivity corresponding to the Dagan model (7.42), and the approximations valid for $\tau \gg 1$, $\tau < 1$, and $\tau \ll 1$. The dimensionless time $\tau = \overline{u}t/\lambda_Y$ equals the number of integral correlation lengths λ_Y that the mean flow travels in time t. The parameters $u_0 = 0.091\ md^{-1}$, $\lambda_Y = 2.8\ m$, $\sigma_Y^2 = 0.24$, $d_L = 10^{-3}\ m^2d^{-1}$, and $d_T = 3 \cdot 10^{-4}\ m^2d^{-1}$ were chosen to simulate the Borden aquifer (Freyberg, 1986). In contrast to the real aquifer, however, we assumed that the log hydraulic conductivity is isotropic.*

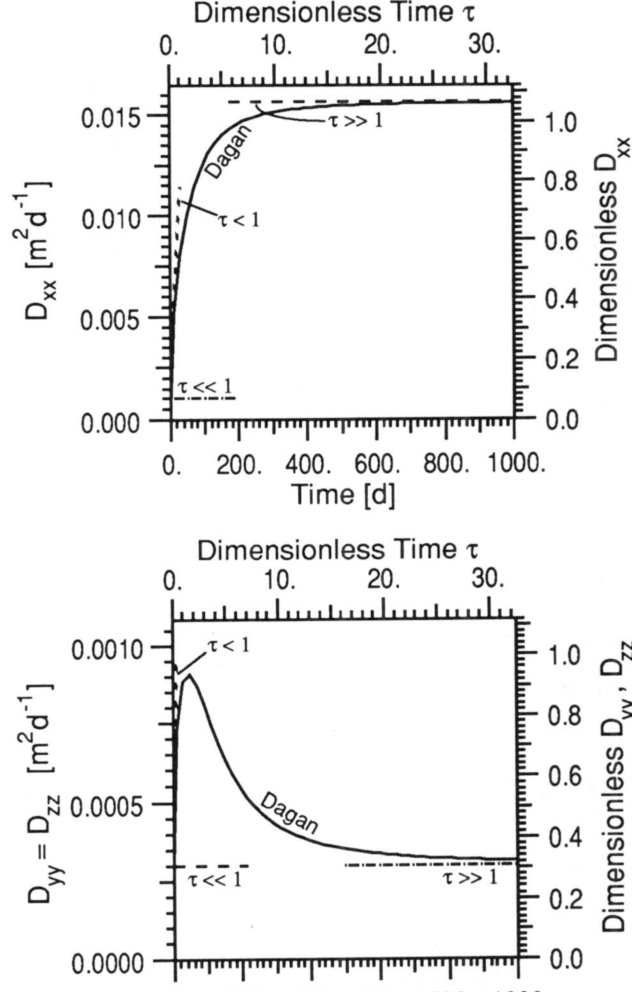

Figure 7.4: Elements of the macrodispersion tensor (7.41) in an aquifer with an isotropic, exponentially decaying covariance (7.42) of the log hydraulic conductivity. The dimensionless time $\tau = \overline{u}t/\lambda_Y$ equals the number of integral correlation lengths λ_Y that the mean flow travels in time t. The dimensionless macrodispersion tensor is obtained by dividing D_{jk} by $\sigma_Y^2 \lambda_Y \overline{u}$. The parameters are the same as in Figure 7.3.

By (7.41) the macrodispersion tensor then becomes

$$\tau < 1: \qquad D_{xx} \approx d_L + \frac{8}{15}\sigma_Y^2 \overline{u}^2 t, \quad D_{yy} \approx d_T + \frac{1}{15}\sigma_Y^2 \overline{u}^2 t \ . \tag{7.48}$$

In this approximation, the spreading of a solute in a stochastic continuum is described by the superposition of (i) a convection-dispersion process (2.58), caused by fluctuations at the microscale, which causes the covariance to increase linearly with time and (ii) a stochastic-convective process (2.68) which causes it to increase quadratically with time. The stochastic-convective process occurs, because for small dimensionless times the aquifer may be approximated by a series of non-interacting stream tubes (see Chapter 4).

For very small dimensionless times, we may even neglect the quadratic term in (7.47) and again obtain a convection-dispersion process. In this approximation, however, we are looking at distances that are smaller than the resolution of the experiment and the "macrodispersion" tensor reduces to the values that we have assumed for the microscale dispersion tensor in the first place. This limit is also obtained when the correlation length of the hydraulic conductivity is assumed to be zero, which is essentially the case if a constant "field-averaged" value of the hydraulic conductivity is used to model solute movement on the field scale.

The spreading of a solute pulse after $\tau = 6.5$ ($t = 200$ d) for the Dagan model (7.43) and for the approximations discussed above is shown in Figure 7.5.

Although the stochastic continuum approach has shown great promise in the description of solute transport in ground water, its assumptions may limit its usefulness in the unsaturated zone. Since flow through the unsaturated zone is normally perpendicular to the direction of natural stratification, the assumption of stationary random soil properties is unlikely to be met in many cases. Further, the connection between solute velocity and hydraulic properties is far more complex in unsaturated soil than in ground water, and the covariance $\boldsymbol{\sigma}_X$ of the particle positions will in general be a complicated function of the covariance structure of several soil hydraulic properties, even under uniform flow. Finally, whereas flow in ground water is normally steady, unsaturated flow is normally transient, so that the solute velocity will be a function of time as well as position.

Despite these limitations, models of this type are being developed now (Mantaglou and Gelhar, 1987abc), but they must overcome formidable obstacles that are not faced in saturated soil. One major difficulty is caused by the nonlinearity of the unsaturated water flow regime. In stochastic continuum analysis, the local solute velocity must be determined, so that its covariance can be calculated. In

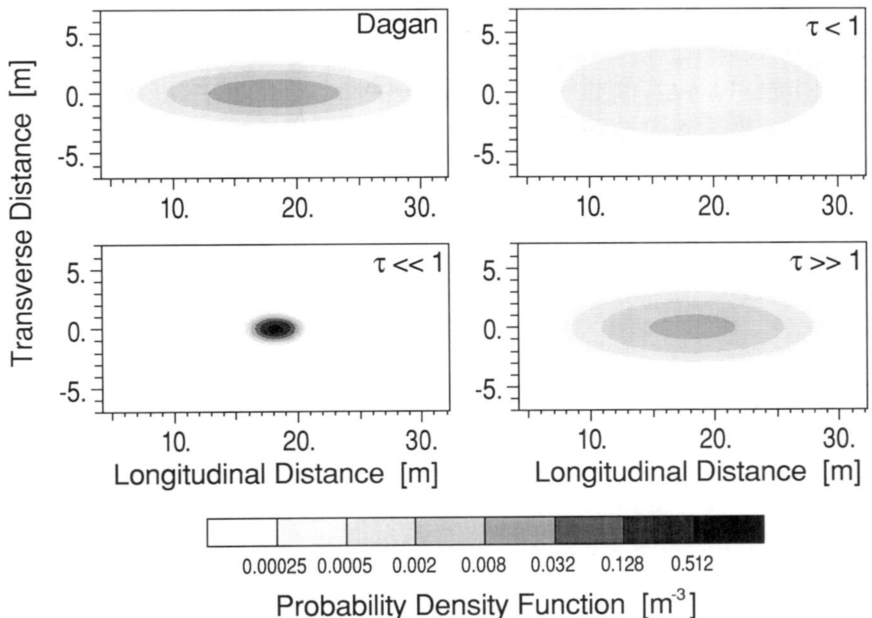

Figure 7.5: *Probability density function $f(x, y, 0, t)$ at dimensionless time $\tau = 6.5$ ($t = 200$ d) for a solute pulse that was present at $(\boldsymbol{x}, t) = (0, 0)$ in the aquifer of Figure 7.3. The pdfs were calculated with (7.38) using the dispersion tensors given by (7.43) and its approximations for the cases $\tau \gg 1$, $\tau < 1$, and $\tau \ll 1$. (Note the logarithmic scale of the gray levels.)*

saturated flow, the solute velocity is related through Darcy's Law (Hillel, 1971) to the saturated hydraulic conductivity, which is observable. However, in unsaturated flow, the solute velocity must be calculated from the Buckingham-Darcy flux law (Hillel, 1971), which involves both the relation $\boldsymbol{K}(\theta)$ between the unsaturated hydraulic conductivity and the water content, and the relation $\psi(\theta)$ between the matric potential and the water content. Therefore, to estimate the spatial structure of the velocity, one must determine the spatial structure of both of these relations over a range of water content and matric potential.

Finally, even if one could predict the spatial structure of the Buckingham-Darcy water flux, the solute velocity may still have an undetermined spatial structure because of the presence of immobile water, preferential flow channels, etc., which involve only a fraction of the wetted pore space. These flow channels are not observable by any current measurement technology except dye tracing, which involves massive excavation, and is only practical with soils of certain colors (Ghodrati, 1989).

For these reasons, the task of developing a continuum description of water and solute flow through unsaturated soil is likely to remain a research problem for the forseeable future. In the interim, advances in computing power have made possible a new generation of solute transport models that describe the motion of individual particles in terms of equations like (7.34), and develop a large scale description of transport by repeating the calculation over thousands of realizations and averaging the results.

7.4 Numerical Simulation of Transport Through a Stochastic Medium

The basic idea of the *particle-tracking* or *random walk* model, that is developed in this section, is to simulate the movement of point-like particles through a stochastic medium. Each trajectory is regarded as one realization of a stochastic process $\boldsymbol{Z}(t)$ that describes the transport of solutes through this medium. Moments of this process are then estimated by averaging over the ensemble of simulated trajectories. In accordance with the formulation of the transfer function in Chapter 2, we will model the mean solute movement which is averaged in the direction perpendicular to the mean flow, and we will only study linear transport; this allows us to restrict the simulations to non-interacting particles which all start at time $t = 0$ from $z = 0$.[9] We assume that $\boldsymbol{Z}(t)$ is a continuous Markov process—i.e. the position of a particle is almost always a continuous function of time, and for times $t > t_0$ (future) it depends only on time t_0 (present), but not on times $t < t_0$ (past)—and will describe it by the pdf $f_Z(z,t)$.[10] One can then show (Gardiner, 1985, Sec. 3.4) that the only permissible changes of $f_Z(z,t)$ during an infinitesimal time dt can be described by a convection velocity V and a dispersion coefficient D, i.e. a delta-function located at z will be transformed into a Gauss function with mean $z + V dt$ and variance $2D dt$. The assumption that the trajectories of the particles are continuous Markov processes therefore leads to a CDE for $f_Z(z,t)$. This may be

[9] In contrast to the previous section, we are now primarily thinking of vertical movement. To indicate this change, we denote the direction of the mean flow with z instead of x as above.

[10] The number $f_Z(z,t)\,dz$ can be interpreted (i) as the probability of finding a specific particle in the interval $z \leq z' < z + dz$ or (ii) as the fraction of all the particles in this interval. Therefore, $f_Z(z,t)$ may be interpreted as the resident concentration at time t corresponding to an initial resident distribution $\delta(z)$ at $t = 0$.

regarded as a justification for using simulations of continuous Markov processes to study solute transport through soil.

Although we want to *model* only horizontally averaged solute transport, we will *simulate* the movement in a two-dimensional stochastic medium where the horizontal coordinate incorporates the spatial variability of the transport parameters on the mesoscale. The medium is thus described locally by the two microscopic parameter fields $v(x,z)$ and $d_d(x,z)$.

To make the simulations more efficient on a computer, we will simulate approximations to the real process which are discrete in time. We therefore assume that the velocity of a particle is constant during the small time interval Δt.[11]

7.4.1 Homogeneous Soil

We will first consider a soil that is homogeneous in both the vertical and the horizontal direction. The convection velocity v and the dispersion coefficient d_d are constant in the entire soil. At time $t = 0$ all the particles are at $z = 0$, corresponding to a Dirac pulse present in the soil. The pdf of the particle positions at time $t > 0$ then corresponds to the resident transfer function (3.57) of an infinite soil. Approximate realizations $Z(t_k) = (x(t_k), z(t_k))$ of the stochastic process $\boldsymbol{Z}(t) = (\boldsymbol{x}(t), \boldsymbol{z}(t))$ at discrete times t_k may be generated with

$$z(t_{k+1}) = z(t_k) + v\Delta t + \omega\sqrt{24 d_d \Delta t}\ , \tag{7.49}$$

and the initial condition

$$z(0) = 0\ , \tag{7.50}$$

where the random variable ω in (7.49) is uniformly distributed in the interval $[-1/2, 1/2]$ and successive realizations are independent. These equations only generate realizations of $\boldsymbol{z}(t)$, but as the soil is homogeneous, the x-coordinate is immaterial. (The factor 24 in front of d_d is introduced to make the dispersion coefficients in (7.49) and (3.57) equal, as will be shown below.) To verify that the pdf of $z(t_k)$ at a fixed time t_k is indeed the Gaussian function (3.57) with mean vt_k and variance $2d_d t_k$, we consider the increment

$$\Delta \boldsymbol{z}_k := \boldsymbol{z}(t_k) - \boldsymbol{z}(t_{k-1}) \tag{7.51}$$

[11] This implies that we are simulating resident concentrations. If we wanted to simulate flux concentrations, we would choose approximations that are discrete in space, i.e. we would assume that the velocity of a particle is constant while it is moving a small distance Δz.

with mean

$$\langle \Delta \boldsymbol{z}_k \rangle = v\Delta t \ , \tag{7.52}$$

and variance

$$\mathrm{Var}(\Delta \boldsymbol{z}_k) = 24 d_d \Delta t \int_{-1/2}^{1/2} \omega^2 \, d\omega = 2 d_d \Delta t \ . \tag{7.53}$$

The position $z(t_k)$ of a particle at time t_k is the sum of k independent and identically distributed increments

$$z(t_k) = \sum_{j=1}^{k} \Delta z_j \ . \tag{7.54}$$

By applying the central limit theorem (Papoulis, 1984) to (7.54), we conclude that for large values of k—which is the case in a typical simulation—the pdf of $\boldsymbol{z}(t_k)$ at time t_k approximates a Gaussian distribution[12]

$$f(z(t_k)) \xrightarrow{k\to\infty} \frac{1}{2\sqrt{\pi d_d t_k}} \exp\Big(-\frac{(z - v t_k)^2}{4 d_d t_k}\Big) \tag{7.55}$$

with mean

$$\langle \boldsymbol{z}(t_k) \rangle = k(v\Delta t) = v t_k \tag{7.56}$$

and variance

$$\mathrm{Var}(\boldsymbol{z}(t_k)) = k(2 d_d \Delta t) = 2 d_d t_k \ . \tag{7.57}$$

7.4.2 Horizontally Inhomogeneous Soil

We will now consider a soil where the microscopic transport parameters are constant in the z-direction (vertical), but vary along the x-direction (horizontal). For simplicity, we describe the horizontal variability by a deterministic periodic function. We assume that the main flow in the soil is vertical, i.e. the convective movement of the particles is along the z-axis. The particles will spread horizontally, however, due to dispersion and will therefore move through regions with different transport parameters.

To simulate trajectories in this soil, we use essentially the same approach as in the homogeneous case. At time $t = 0$ all the particles are at $z = 0$, evenly distributed in the horizontal direction. The convection velocity $v(x)$ and the vertical dispersion coefficient $d_z(x)$ depend on the horizontal position

[12] The Gaussian distribution of the sum is independent of the distribution of the increments. For this reason the distribution of ω in (7.49) is immaterial and was chosen to achieve maximum efficiency on a computer.

of a particle, whereas the horizontal dispersion coefficient d_x is assumed to be constant. Approximate realizations $(x(t_k), z(t_k))$ of the particle trajectories are generated in analogy to (7.49) by

$$\begin{aligned} x(t_{k+1}) &= x(t_k) + \omega\sqrt{24 d_x \Delta t} \ , \\ z(t_{k+1}) &= z(t_k) + v(x)\Delta t + \omega\sqrt{24 d_z(x)\Delta t} \ . \end{aligned} \tag{7.58}$$

As the horizontal variability was assumed to be periodic, we will consider only a single period and, for convenience, rescale and shift the x-coordinate so that a period covers the interval $[-1/2, 1/2]$. The periodicity of the horizontal variability requires $v(-1/2) = v(1/2)$ and $d_z(-1/2) = d_z(1/2)$.

The variability of v and d_z in the horizontal direction, together with the horizontal dispersion, causes a correlation between the increments (7.51) for times $t > \Delta t$, which makes the calculation of travel depth moments much more difficult. Two limiting cases—$d_x = 0$ and $d_x \to \infty$—can be solved easily, however.

No Horizontal Dispersion ($d_x = 0$)

For $d_x = 0$ the soil may be regarded as a collection of non-interacting, homogeneous soil columns each of which is described by a CDE (see Chapter 4) and the pdf of the particle positions in the entire medium may formally be written as

$$f(z, t_k) \xrightarrow{k\to\infty} \int_{-1/2}^{+1/2} f(z, t_k; v(x), d_z(x))\, dx \ , \tag{7.59}$$

where $f(z, t_k; v(x), d_z(x))$ is the pdf (7.55) for a homogeneous soil with parameters $v(x)$ and $d_z(x)$. In general, (7.59) can only be evaluated numerically. However, the moments $M_N(t_k)$ of $f(z, t_k)$ may often be calculated analytically in terms of the local moments $m_N(t_k; x)$ of (7.55) as

$$\begin{aligned} M_N(t_k) &:= \int_{-\infty}^{+\infty} z^N f(z, t_k)\, dz \\ &\xrightarrow{k\to\infty} \int_{-\infty}^{+\infty} z^N \int_{-1/2}^{+1/2} f(z, t_k; v(x), d_z(x))\, dx dz \\ &= \int_{-1/2}^{+1/2} m_N(t_k; x)\, dx \ . \end{aligned} \tag{7.60}$$

The mean and variance of the particle position at time t_k may be calculated from (7.60) as (see Problem 7.6)

$$\langle z(t_k)\rangle = \langle v(x)\rangle t_k \,, \tag{7.61}$$

$$\begin{aligned} \mathrm{Var}(z(t_k)) &= M_2(t_k) - M_1^2(t_k) \\ &= 2\Big[\langle d_z(x)\rangle + \frac{t_k}{2}\,\mathrm{Var}(v)\Big]t_k \,, \end{aligned} \tag{7.62}$$

where the mean values of v and d_z are defined as

$$\begin{aligned} \langle v(x)\rangle &:= \int_{-1/2}^{1/2} v(x)\,dx \\ \langle d_z(x)\rangle &:= \int_{-1/2}^{1/2} d_z(x)\,dx \end{aligned} \tag{7.63}$$

and the variance of v as

$$\mathrm{Var}(v) := \langle v^2(x)\rangle - \langle v(x)\rangle^2 \,. \tag{7.64}$$

From (7.61)–(7.62) we may define the macroscopic parameters V and D_z of the entire medium as

$$V := \langle v(x)\rangle \,, \tag{7.65}$$

$$D_z := \langle d_z(x)\rangle + \frac{t_k}{2}\,\mathrm{Var}(v) \,. \tag{7.66}$$

These parameters describe a homogeneous medium whose first two moments are identical to those of the inhomogeneous medium.

We notice from (7.66) that the entire medium is no longer described by a CDE because the macrodispersion coefficient D_z is growing linearly with time.

Example 7.2 DOUBLE POROUS MEDIUM WITHOUT HORIZONTAL DISPERSION

A double porous medium is the sum of two homogeneous media with different transport parameters. Without horizontal dispersion, $d_x = 0$, the particles cannot move from one medium into the other. The motion of each particle is therefore governed by constant parameters, whose values depend on the x-coordinate of the particle at $t = 0$.

We denote the volume fraction of medium 1 by ϑ_1 and its transport parameters with v_1 and $d_{z,1}$. Correspondingly, the volume fraction and the transport parameters in medium 2 are $1-\vartheta_1$, v_2, and $d_{z,2}$. The convection velocity on the interval $[-1/2, 1/2]$ may thus be written as the symmetric function

$$v(x) = \begin{cases} v_1 & ;\ |x| \le \vartheta_1/2 \\ v_2 & ;\ |x| > \vartheta_1/2 \end{cases} \tag{7.67}$$

and the dispersion coefficient in the vertical direction as

$$d_z(x) = \begin{cases} d_{z,1} & ; |x| \leq \vartheta_1/2 \\ d_{z,2} & ; |x| > \vartheta_1/2 \end{cases} . \tag{7.68}$$

We may calculate the macroscopic parameters of the entire medium with (7.63)–(7.66) and obtain

$$\begin{aligned} V &= \int_{-1/2}^{1/2} v(x)\,dx = 2\Big(\int_0^{\vartheta_1/2} v_1\,dx + \int_{\vartheta_1/2}^{1/2} v_2\,dx\Big) \\ &= \vartheta_1 v_1 + (1-\vartheta_1)v_2 , \end{aligned} \tag{7.69}$$

$$\begin{aligned} D_z(t_k) &= \int_{-1/2}^{1/2} d_z(x) t_k\,dx + \frac{t_k}{2}\,\mathrm{Var}(v) \\ &= \vartheta_1 d_{z,1} + (1-\vartheta_1)d_{z,2} + \frac{t_k}{2}\Big[\int_{-1/2}^{1/2} v^2(x)\,dx - \Big(\int_{-1/2}^{1/2} v(x)\,dx\Big)^2\Big] \\ &= \vartheta_1 d_{z,1} + (1-\vartheta_1)d_{z,2} + \frac{t_k}{2}\vartheta_1(1-\vartheta_1)(v_1-v_2)^2 . \end{aligned} \tag{7.70}$$

We notice that although the first two moments of the travel depth pdf of the homogeneous medium described by (7.69)–(7.70) and of the double porous medium are identical, the two pdfs are not similar at all. The distribution of the travel depths for the double porous medium is bimodal—it is the sum of two Gaussian distributions—whereas it is unimodal for the homogeneous medium.

Infinite Horizontal Dispersion ($d_x \to \infty$)

For $d_x \to \infty$ the horizontal position of a particle at time t_2 is independent from its position at t_1 if $|t_2 - t_1| > \Delta t$ and its pdf is constant in the interval $[-1/2, 1/2]$. This implies that the parameters $v(x)$ and $d_z(x)$ which determine the motion of the particle are independent for times t_2 and t_1. Therefore, the increments (7.51) are independent. This allows us to calculate the travel depth pdf $f(z(t_k))$ for large times t_k with the central limit theorem. We obtain

$$f(z(t_k)) \xrightarrow{k\to\infty} \frac{1}{2\sqrt{\pi D_z t_k}} \exp\Big(-\frac{(z-Vt_k)^2}{4D_z t_k}\Big) , \tag{7.71}$$

where

$$V := \langle v(x) \rangle , \tag{7.72}$$

$$D_z := \langle d_z(x) \rangle + \frac{\Delta t}{2}\,\mathrm{Var}(v) \tag{7.73}$$

are the macroscopic parameters of the entire medium (see Problem 7.6)

In contrast to the case without horizontal dispersion—where the vertical macrodispersion coefficient grows linearly with time—the vertical macrodispersion coefficient D_z is now constant and approximates the average local dispersion coefficient $\langle d_z(x)\rangle$ as $\Delta t \to 0$. Equation (7.73) may be used to determine the time step Δt of the simulation in terms of the local parameters v and d_z, and of the precision $\Delta D_z := D_z - \langle d_z(x)\rangle$ that is desired,

$$\Delta t = \frac{2\Delta D_z}{\mathrm{Var}(v)} \ . \tag{7.74}$$

Intermediate Horizontal Dispersion ($0 < d_x < \infty$)

Equation (7.58) may be solved numerically for $0 < d_x < \infty$. To do this, we have to assume a specific functional form for the microscopic transport parameters $v(x)$ and $d_z(x)$. For illustration, we will choose the two parameters to be perfectly correlated

$$\begin{aligned} v(x) &= v_0 + \Delta v g(x) \ , \\ d_z(x) &= d_{z,0} + \Delta d_z g(x) \ , \end{aligned} \tag{7.75}$$

where the function

$$g(x) = \frac{1}{1+(50x)^2} \ , \tag{7.76}$$

which has a rather sharp peak at $x = 0$, is intended to simulate a soil where preferential flow occurs (Roth, 1989). To calculate the macroscopic parameters V and D_z, we first estimate the moments of the particle positions by

$$M_n'(t_k) = \frac{1}{N}\sum_{j=1}^{N} z^n(t_k) \ , \tag{7.77}$$

where N is the number of particles that are used in the simulation. The parameters are then obtained by

$$\begin{aligned} V(t_k) &= \frac{M_1'(t_{k+1}) - M_1'(t_{k-1})}{2\Delta t} \ , \\ D_z(t_k) &= \frac{\mathrm{Var}_z(t_{k+1}) - \mathrm{Var}(t_{k-1})}{2\Delta t} \ , \end{aligned} \tag{7.78}$$

where

$$\mathrm{Var}_z(t_k) = M_2'(t_k) - M_1'^2(t_k) \ . \tag{7.79}$$

A plot of the macrodispersion coefficient D_z as a function of time for different values of the horizontal dispersion coefficient d_x is shown in Figure 7.6. We obtain the linear growth of D_z with time (7.66) for $d_x = 0$, and a very small constant value (7.73) for $d_x \to \infty$. For intermediate values of d_x, the macrodispersion first increases linearly with time and approaches a constant value for large times. The asymptotic value and the time scale at which it is reached depend on the time $\tau := \lambda_h^2/(2d_x)$, which may be thought of as the characteristic time required for a particle to diffuse horizontally through one integral correlation length λ_h.

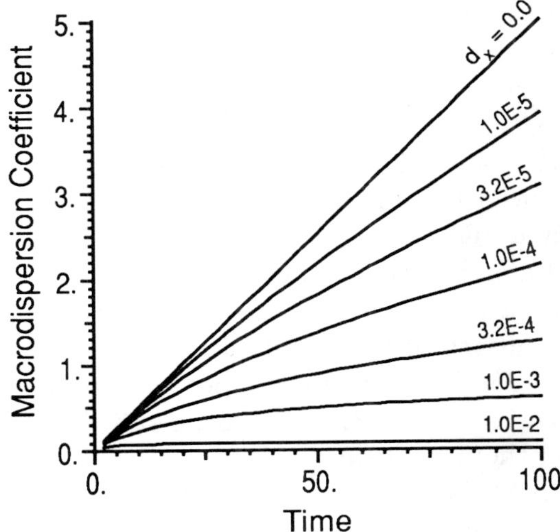

Figure 7.6: Vertical dispersion coefficient $D_z(t)$ for different values of the horizontal dispersion coefficient d_x. The graph was obtained from the simulation of the movement of 2^{16} particles with the following parameters: $\lambda_h = 1$, $\langle d_z(x) \rangle = 0.0216$, $\mathrm{Var}(v) = 0.0999$, $\Delta t = 0.0400$.

Comparing Figures 7.1, 7.4 and 7.6, we notice the striking similarity. All of the dispersion representations essentially show the spreading of particles as they move through a spatially correlated, heterogeneous system. For distances from the point of solute entry that are much smaller than the correlation length of the system, the dispersion coefficient increases linearly, and it approaches a constant value at distances much larger than the correlation length.

In this chapter we have studied several approaches for extending our understanding of solute transport from the macroscale to the mesoscale. The desire to do this arises—apart from scientific curiosity—from the hope to find an alternative to measuring the transfer function for the entire transport volume of interest. Whereas it might be feasible to measure solute outflow concentrations in the surface soil zone, experimental determination of the impulse response function of an aquifer or deep unsaturated zone may not be, particularly if the characteristic flow rate is small. Yet there are many practical situations, e.g. waste spills, where the ultimate goal of the model is to predict solute migration at this large scale. One alternative discussed throughout this book is to conduct a small-scale measurement of the transfer function parameters and extrapolate to larger scales with a process model. As we have seen, the local measurements may be extrapolated only for homogeneous soil, and further must rely on zero or infinte time limits if the correlation structure of the travel times is not known. Neither limit is reasonable for many practical applications.

These problems could be solved, at least in stationary media, if we had the ability to integrate measurements of local (mesoscale) properties to obtain a macroscopic description instead of letting nature do it and looking at the emerging solution. The ideas presented in this chapter indicate that a prerequisite for this integration is the knowledge of the covariance structure of the parameter fields that determine transport on the mesoscale. Although it has proven feasible to generate a sufficient number of measurements of saturated hydraulic conductivity to calculate the covariance structure of the velocity field of a stationary aquifer (Freyberg, 1986), the unsaturated soil regime will prove to be a more formidable obstacle. Since water velocity depends on water content in unsaturated soil, simultaneous local measurements of unsaturated hydraulic conductivity and matric potential over a wide range of saturation must be generated at a spatial resolution sufficient to analyse for the correlation structure on the mesoscale.

At the moment there seem to be two equally uneasy alternatives for determining solute transport on a large scale: spending one's time in looking at the movement of a tracer pulse that will be analysed by transfer functions, or spending one's time and money in doing lots of small scale experiments and integrating them to a macroscopic picture.

Problems

Problem 7.1 Show that the variance (7.3) reduces to (7.5) when the correlation coefficient is given by (7.4).

Problem 7.2 Calculate the asymptotic values of the generalized dispersivity λ (7.9) for large and small z.

Problem 7.3 Derive (7.27), relating the travel time variance to the autocorrelation function of the fluctuation of the inverse velocity.

Problem 7.4 Prove that the multivariate Gaussian pdf (7.38) for the displacement is a solution of the generalized CDE (7.39).

Problem 7.5 Spatial correlations may be generated with the simple autoregressive AR-1 model (Box and Jenkins, 1978)

$$Z_j - m = \rho(Z_{j-1} - m) + s\xi\sqrt{1-\rho^2} \; , \tag{7.80}$$

where $Z_j := Z(X_j)$ is a weakly stationary random function which has mean m and variance s^2 at each position X_j, ρ is the correlation coefficient, and $\xi_j := \xi(X_j)$ is a spatially uncorrelated, normally distributed random function with zero mean and unit variance. Calculate the variogram and the autocorrelation function for this model. What is the integral length scale?

Problem 7.6 Derive the macroscopic parameters V and D_z for the two cases $d_x = 0$ ((7.65)–(7.66)) and $d_x \to \infty$ ((7.72)–(7.73)).

Appendix A

Integral Transforms

An integral transform with respect to a given variable is a mathematical operation on a function of one or more variables that removes the dependence of the function on the given variable. This operation consists of a definite integration of the function, multiplied by a weighting factor, over the entire domain of the variable. The weighting factor contains the variable, and a parameter, which remains as part of the transformed function.

The principal use of integral transforms is to simplify partial differential equations by transforming them to remove one of their variables, thereby allowing them to be solved in the transformed space. The solution is then transformed back to the original space by the inverse transformation operation.

We employ two different transformation methods in this book. The *Laplace transformation* is used to transform variables, such as the time t, which are defined for all non-negative values of their arguments ($t \in [0, \infty]$). This allows the Laplace transformation to be used to transform both the differential equations for transport processes, and their boundary conditions. The Laplace transform may also be used to transform the space variable z in applications of the transport theory where z is semi-infinite.

The *Fourier transform* is used to transform a variable which is defined for all values ($z \in [-\infty, \infty]$) of its argument. Therefore, this transform will be used to remove the space dependence of a differential equation defined over the entire real domain.

A.1 The Laplace Transform

The Laplace transform $\widehat{f}(s)$ of a function $f(t)$ is defined as

$$\widehat{f}(s) \equiv \mathcal{L}(f(t)) := \int_0^\infty f(t) \exp(-st)\, dt \tag{A.1}$$

for $0 \leq t < \infty$ and for values of the conjugate variable s where the integral exists. The definition (A.1) can be extended to a function $f(x_1, \ldots, x_n, t)$ of several variables,

$$\hat{f}(x_1, \ldots, x_n; s) := \int_0^\infty f(x_1, \ldots, x_n, t) \exp(-st)\, dt \ . \tag{A.2}$$

Example A.1 ANALYTIC EVALUATION OF LAPLACE TRANSFORMS

Many functions (or generalized functions) have Laplace transforms which can be obtained by direct integration of (A.1). Examples of several simple transforms are given below.

- $f(t) = H(t)$

$$\hat{f}(s) = \int_0^\infty \exp(-st)\, dt = -\frac{1}{s} \exp(-st)\Big|_{t=0}^{t=\infty} = \frac{1}{s} \tag{A.3}$$

- $f(t) = \delta(t-a)\ ; \quad a \geq 0$

$$\hat{f}(s) = \int_0^\infty \delta(t-a) \exp(-st)\, dt = \exp(-sa) \tag{A.4}$$

A special case of (A.4) for $a = 0$ is $\mathcal{L}\big(\delta(t)\big) = 1$.

- $f(t) = t^N$

$$\hat{f}(s) = \int_0^\infty t^N \exp(-st)\, dt = \frac{1}{s^{N+1}} \int_0^\infty y^N \exp(-y)\, dy = \frac{N!}{s^{N+1}} \tag{A.5}$$

A.1.1 Transforms of Derivatives and Integrals

One of the major uses of Laplace transforms is in the solution of ordinary or partial differential equations, where often the dependent variable is differentiated or integrated with respect to time. The Laplace transform operation (A.1) of multiplying by $\exp(-st)$ and integrating over t from 0 to ∞ can be applied to each side of a differential equation, and to the boundary conditions. As a preface to that exercise, the Laplace transform of a derivative and integral of a function with respect to t will be evaluated. In these exercises, as well as in future calculations, extensive use will be made of the formula for integrating by parts,

$$\int_a^b u(x)\, dv(x) = u(x)\, v(x)\Big|_a^b - \int_a^b v(x)\, du(x) \ . \tag{A.6}$$

- $f(t) = \dfrac{\partial y(x,t)}{\partial t}$

$$\begin{aligned}\widehat{f}(s) &= \int_0^\infty \frac{\partial y(x,t)}{\partial t} \exp(-st)\,dt \\ &= y(x,t)\exp(-st)\Big|_0^\infty + s\int_0^\infty y(x,t)\exp(-st)\,dt\ ,\end{aligned} \tag{A.7}$$

or,

$$\mathcal{L}\Big(\frac{\partial y(x,t)}{\partial t}\Big) = -y(x,0) + s\widehat{y}(x;s)\ . \tag{A.8}$$

Note that the initial value $y(x,0)$ of the function appears as part of the transform of the time derivative of $y(x,t)$. The Laplace transform of the second time derivative is similar, except that two integrations by parts must be carried out.

- $f(t) = \dfrac{\partial^2 y(x,t)}{\partial t^2}$

$$\begin{aligned}\widehat{f}(s) &= \int_0^\infty \frac{\partial^2 y(x,t)}{\partial t^2} \exp(-st)\,dt \\ &= \frac{\partial y(x,t)}{\partial t}\exp(-st)\Big|_0^\infty + s\int_0^\infty \frac{\partial y(x,t)}{\partial t}\exp(-st)\,dt\ .\end{aligned} \tag{A.9}$$

With (A.8), (A.9) may be written

$$\mathcal{L}\Big(\frac{\partial^2 y(x,t)}{\partial t^2}\Big) = -\frac{\partial y}{\partial t}(x,0) - sy(x,0) + s^2\widehat{y}(x;s)\ . \tag{A.10}$$

If the function is differentiated by a variable other than the one being transformed, the order of differentiation and integration may be reversed.

- $f(t) = \dfrac{\partial^N y(x,t)}{\partial x^N}$

$$\widehat{f}(s) = \int_0^\infty \frac{\partial^N y(x,t)}{\partial x^N}\exp(-st)\,dt = \frac{\partial^N}{\partial x^N}\int_0^\infty y(x,t)\exp(-st)\,dt\ . \tag{A.11}$$

Therefore,

$$\mathcal{L}\Big(\frac{\partial^N y(x,t)}{\partial x^N}\Big) = \frac{d^N\widehat{y}(x;s)}{dx^N}\ . \tag{A.12}$$

Note that if we regard s as a parameter, then the derivative of the Laplace transform in (A.12) may be written as a total derivative rather than as a partial one.

The integral of a function over t may be evaluated with the integration by parts formula (A.6)

- $f(t) = \int_0^t y(x,t')\,dt'$

$$\begin{aligned}\widehat{f}(s) &= \int_0^\infty \int_0^t y(x,t')\,\exp(-st)\,dt'\,dt \\ &= \left(-\frac{1}{s}\exp(-st)\int_0^t y(x,t')\,dt'\right)\Bigg|_{t=0}^{t=\infty} + \frac{1}{s}\int_0^\infty y(x,t)\,\exp(-st)\,dt\ ,\end{aligned} \tag{A.13}$$

or,

$$\mathcal{L}\Big(\int_0^t y(x,t')\,dt'\Big) = \frac{1}{s}\widehat{y}(x;s)\ . \tag{A.14}$$

A.1.2 Transformed Solution of Differential Equations

The formulas given in (A.7)–(A.14) make it possible to calculate the Laplace transform of the solution to many ordinary and partial differential equations of interest in physics and engineering, as illustrated in the next two examples.

Example A.2 LAPLACE TRANSFORM SOLUTION OF THE DAMPED HARMONIC OSCILLATOR

A spring of force constant k is attached at one end to a mass m lying on a table, and at the other end to a wall at the end of the table. The equilibrium length of the spring (at which it exerts no force on the mass) is x_0. At time $t = 0$, the spring is stretched out to a position $x_1 > x_0$ and released. We wish to calculate the motion $x(t)$ of the mass for $t > 0$, assuming that the table exerts a frictional force $F = -\beta dx(t)/dt$ on the mass.

By Newton's first law of motion,

$$\sum_i F_i = ma = m\frac{d^2x}{dt^2} = -\beta\frac{dx}{dt} - k(x - x_0)\ , \tag{A.15}$$

where F_i are the forces (friction and spring) on the mass. Equation (A.15) is to be solved subject to the initial conditions

$$x(0) = x_1\ , \tag{A.16}$$

$$\frac{dx}{dt}(0) = 0\ . \tag{A.17}$$

The Laplace transform of (A.15) may be calculated using (A.3), (A.7), (A.9), (A.16) and (A.17). The result is

$$ms^2\widehat{x} - msx_1 + \beta s\widehat{x} - \beta x_1 + k\widehat{x} - \frac{kx_0}{s} = 0\ . \tag{A.18}$$

This is merely an algebraic equation for the function $\widehat{x}(s)$, the transform of $x(t)$. Thus

$$\widehat{x}(s) = \frac{msx_1 + \beta x_1 + kx_0/s}{ms^2 + \beta s + k}\ , \tag{A.19}$$

which is the Laplace transform of the solution to (A.15)–(A.17).[1]

Example A.3 LAPLACE TRANSFORM SOLUTION OF THE HEAT EQUATION

When one-dimensional partial differential equations involving a space and time variable are Laplace-transformed, they become ordinary differential equations. This will be illustrated on the heat flow equation, for the case of a semi-infinite medium ($0 < x < \infty$) at zero initial temperature, whose inlet surface is held at the constant value T_0 for all $t > 0$. The heat transport equation is (Carslaw and Jaeger, 1959)

$$\frac{\partial T(x,t)}{\partial t} = K_T \frac{\partial^2 T(x,t)}{\partial x^2} \tag{A.20}$$

with initial and boundary conditions

$$T(x,0) = 0 \; , \tag{A.21}$$

$$T(\infty,t) = 0 \; , \tag{A.22}$$

$$T(0,t) = T_0 \; , \tag{A.23}$$

where K_T is the thermal diffusivity.

After Laplace transformation, (A.20)–(A.23) may be written as

$$\frac{d^2\widehat{T}(x)}{dx^2} - q^2\widehat{T}(x) = 0 \; , \tag{A.24}$$

$$\widehat{T}(\infty) = 0 \; , \tag{A.25}$$

$$\widehat{T}(0) = \frac{T_0}{s} \; , \tag{A.26}$$

where $q := \sqrt{s/K_T}$. If we substitute the trial solution

$$\widehat{T}(x) = \exp(-mx) \tag{A.27}$$

(A.24) becomes

$$(m^2 - q^2)\exp(-mx) = 0 \; , \tag{A.28}$$

which for arbitrary x is valid only if $m = \pm q$. Therefore, the general solution to (A.24) can be written

$$\widehat{T}(x) = A\exp(-qx) + B\exp(qx) \; , \tag{A.29}$$

where A and B are constants which may depend on the variable s but not on x. Since $q > 0$, (A.25) can be valid only if $B = 0$. Then, by (A.26)

$$\widehat{T}(0) = A = \frac{T_0}{s} \longrightarrow \widehat{T}(x) = \frac{T_0}{s}\exp(-qx) \; , \tag{A.30}$$

which is the Laplace transform of the solution to (A.20)–(A.23).

[1]Interested readers may evaluate the inverse transform of (A.19) using the table of transforms in Appendix C.

A.1.3 Inverse Laplace Transformation

Direct Integration

The formal mathematical operation by which $f(t)$ is calculated from $\widehat{f}(s)$ is called a Bromwich contour integral (Carslaw and Jaeger, 1959), written formally as

$$f(t) = \mathcal{L}^{-1}(\widehat{f}(s)) := \frac{1}{2\pi i} \int_{\gamma - i\infty}^{\gamma + i\infty} \widehat{f}(s) \exp(st)\, ds \ , \tag{A.31}$$

where γ is a positive constant, and $i := \sqrt{-1}$. The contour of the integral is chosen by selecting γ large enough so that all singularities of the function $\widehat{f}(s)$ lie to the left of the line drawn normal to the real axis through the point γ. The integral is then evaluated using the calculus of residues (Arfken, 1985).

The contour integration (A.31) is often rather cumbersome, and will not be dealt with further in the book. There are two other alternatives for the inversion operation (numerical inversion and the inversion tables), which will allow many transformed solutions to transport problems to be inverted.

Numerical Inversion

Numerical evaluation of the Bromwich integral (A.31) is possible in many cases if the function $f(t)$ is reasonably smooth. A Fortran program for a versatile inversion method introduced by Talbot (1980) is provided and illustrated in Appendix B. Numerical inversion is often computationally more efficient than direct evaluation of the analytic expressions found in inversion tables, or by contour integration.

Inversion Tables

The most common method of inversion is to use the tables of standard forms furnished in numerous mathematical handbooks. A set of the most commonly encountered transforms in solute transport is provided in Appendix C.

For example, the inverse transform of the solution (A.30) to the heat flow problem in Example A.3 is given by (C.9) of Appendix C

$$T(x,t) = T_0 \,\mathrm{erfc}\Big(\frac{x}{2\sqrt{K_T t}}\Big) \ . \tag{A.32}$$

In many cases, a Laplace transform which is not in the table can be converted to one of the standard forms by various transform operations. The most useful of these are given below.

Laplace Transform Operations

- Shifting Operation

$$\hat{f}(s) = \mathcal{L}(f(t))$$
$$\Downarrow \tag{A.33}$$
$$\mathcal{L}^{-1}(\hat{f}(as+b)) = \exp\Big(-\frac{bt}{a}\Big)\mathcal{L}^{-1}(\hat{f}(as)) = \frac{1}{a}\exp\Big(-\frac{bt}{a}\Big)f\Big(\frac{t}{a}\Big)$$

This operation has many applications, as shown in the next example.

Example A.4 Inversion of the Transform of the Flux PDF of the CDE

The Laplace transform of the CDE flux pdf (2.50) was calculated in Example 2.2. This function does not appear in the table of transforms. However, using (A.33), we may transform it in the following way. First, we write (2.50) as

$$\begin{aligned}
\hat{f}^f(z;s) &= \exp\left[\frac{Vz}{2D}\Big(1-\sqrt{1+\frac{4sD}{V^2}}\Big)\right] \\
&= \exp\Big(\frac{Vz}{2D}\Big)\exp\Big(-\frac{Vz}{2D}\sqrt{\frac{4D}{V^2}}\sqrt{\frac{V^2}{4D}+s}\Big) \\
&= \exp\Big(\frac{Vz}{2D}\Big)\exp\Big(-\frac{z}{\sqrt{D}}\sqrt{\frac{V^2}{4D}+s}\Big) .
\end{aligned} \tag{A.34}$$

This expression contains s only in the form $s + V^2/4D$. Therefore, the inverse transform can be transformed by (A.33) to (letting $x = z/\sqrt{D}$)

$$f^f(z,t) = \mathcal{L}^{-1}\big(\hat{f}^f(z;s)\big) = \exp\Big(\frac{Vz}{2D} - \frac{V^2t}{4D}\Big)\mathcal{L}^{-1}\big(\exp(-x\sqrt{s})\big) . \tag{A.35}$$

We may now complete the inversion with the aid of (C.7) of the Table of Transforms, producing

$$\begin{aligned}
f^f(z,t) &= \exp\Big(\frac{Vz}{2D} - \frac{V^2t}{4D}\Big)\frac{z}{2\sqrt{\pi D t^3}}\exp\Big(-\frac{z^2}{4Dt}\Big) \\
&= \frac{z}{2\sqrt{\pi D t^3}}\exp\Big(-\frac{(z-Vt)^2}{4Dt}\Big) ,
\end{aligned} \tag{A.36}$$

which is the expression (2.51) for the flux pdf of the CDE.

- Convolution Operation

$$\hat{f}_1(s) = \mathcal{L}(f_1(t)) \quad \text{and} \quad \hat{f}_2(s) = \mathcal{L}(f_2(t))$$
$$\Downarrow \tag{A.37}$$
$$\hat{f}_1(s)\hat{f}_2(s) = \mathcal{L}\Big(\int_0^t f_1(\tau)f_2(t-\tau)\,d\tau\Big) = \mathcal{L}\Big(\int_0^t f_1(t-\tau)f_2(\tau)\,d\tau\Big)$$

The convolution operation will allow many new functions to be evaluated from a smaller table of transforms, as shown in the next example.

Example A.5 INVERSION OF THE TRANSFORM OF THE FLUX CDF OF THE CDE

The transfer function equation (2.4) is a convolution integral. Therefore,

$$\widehat{C}^f(z;s) = \widehat{C}^f(0;s)\widehat{f}^f(z;s) \ . \tag{A.38}$$

For example, the step function condition transforms into

$$C^f(0,t) = H(t) \Rightarrow \widehat{C}^f(0;s) = \frac{1}{s} \ . \tag{A.39}$$

Therefore, by (A.37) the inverse transform of (A.38)–(A.39) is

$$\mathcal{L}^{-1}\Big(\frac{1}{s}\exp(-qz)\Big) = \int_0^t \frac{z}{2\sqrt{\pi D\tau^3}}\exp\Big(-\frac{(z-V\tau)^2}{4D\tau}\Big)\,d\tau \ . \tag{A.40}$$

However, we may also use the shifting operation (A.33) on the inverse transform in (A.40), producing

$$\begin{aligned}\mathcal{L}^{-1}\Big(\frac{1}{s}\exp(-qz)\Big) &= \exp\Big(\frac{Vz}{2D}-\frac{V^2t}{4D}\Big)\mathcal{L}^{-1}\Big(\frac{1}{s-V^2/4D}\exp(-x\sqrt{s})\Big) \\ &= \frac{1}{2}\Big[\operatorname{erfc}\Big(\frac{z-Vt}{2\sqrt{Dt}}\Big)+\exp\Big(\frac{Vz}{D}\Big)\operatorname{erfc}\Big(\frac{z+Vt}{2\sqrt{Dt}}\Big)\Big]\end{aligned} \tag{A.41}$$

by (C.11). Thus, the Laplace transform has also provided an indirect means of evaluating a difficult integral, namely

$$\begin{aligned}&\int_0^t \frac{z}{2\sqrt{\pi D\tau^3}}\exp\Big(-\frac{(z-V\tau)^2}{4D\tau}\Big)\,d\tau \\ &= \frac{1}{2}\Big[\operatorname{erfc}\Big(\frac{z-Vt}{2\sqrt{Dt}}\Big)+\exp\Big(\frac{Vz}{D}\Big)\operatorname{erfc}\Big(\frac{z+Vt}{2\sqrt{Dt}}\Big)\Big] \ .\end{aligned} \tag{A.42}$$

- Generalized Convolution Integral (Walker, 1987)

$$\begin{gathered} f(t_1,t_2) = \mathcal{L}_{s_1}^{-1}\Big(\mathcal{L}_{s_2}^{-1}(g(s_1,s_2))\Big) \\ \Downarrow \\ \int_0^t f(\tau,t-\tau)\,d\tau = \mathcal{L}^{-1}(g(s,s)) \ . \end{gathered} \tag{A.43}$$

Equation (A.43) is an extremely valuable tool for inverting complex transforms, by allowing the s-dependence of the transform to be split up into two terms which are inverted separately. This procedure will be illustrated in the next example.

Example A.6 Inversion of the Flux PDF of the CDE Under Rate Limited Adsorption

The flux pdf of the CDE under rate limited adsorption was solved in Example 4.4 as a Laplace transform. Equation (4.30) can be written as (using (4.24) and (4.28))

$$\widehat{C}^f(z;s_1,s_2) = \exp\Big(\frac{Vz}{2D}\Big)\exp\Big(-\frac{Vz}{2D}\sqrt{1+\frac{4D}{V^2}\Big(s_1+\frac{s_2\beta(R-1)}{s_2+\beta}\Big)}\Big)\,, \tag{A.44}$$

where $\beta := \alpha/\rho_b$. In the first inversion with respect to s_1, the parameter s_2 may be treated as a constant. Therefore, we may use the shifting theorem (A.33) to produce

$$\begin{aligned} &\mathcal{L}_{s_1}^{-1}\big(\widehat{C}^f(z;s_1,s_2)\big) \\ &= \exp\Big(-\frac{s_2\beta(R-1)t_1}{s_2+\beta}\Big)\,\mathcal{L}_{s_1}^{-1}\Big[\exp\Big[\frac{Vz}{2D}\Big(1-\sqrt{1+\frac{4Ds_1}{V^2}}\Big)\Big]\Big] \\ &= \exp\Big(-\frac{s_2\beta(R-1)t_1}{s_2+\beta}\Big)\,f^f(z,t_1) \\ &= \exp(-\beta(R-1)t_1)\,\exp\Big(\frac{\beta^2(R-1)t_1}{s_2+\beta}\Big)\,f^f(z,t_1)\,, \end{aligned} \tag{A.45}$$

where $f^f(z,t_1)$ is the flux pdf (2.51) of a mobile, nonadsorbing chemical expressed in terms of t_1 rather than t. The second inversion with respect to s_2 involves only the middle term, whose inverse transform is given as (C.25) in Appendix C. Thus

$$\begin{aligned} f(t_1,t_2) &= \mathcal{L}_{s_2}^{-1}\mathcal{L}_{s_1}^{-1}\big(\widehat{C}^f(z;s_1,s_2)\big) \\ &= \exp(-\beta(R-1)t_1)\,f^f(z,t_1) \\ &\quad \Big[\delta(t_2)+I_1\Big(2\beta\sqrt{(R-1)t_1t_2}\Big)\sqrt{\frac{\beta^2(R-1)t_1}{t_2}}\,\exp(-\beta t_2)\Big]\,, \end{aligned} \tag{A.46}$$

where I_1 is a modified Bessel function (Abramowitz and Stegun, 1970). Thus, the analytic solution for the flux concentration of the CDE with rate limited adsorption is given by applying (A.43) to (A.46)

$$\begin{aligned} C^f(z,t) &= \exp(-\beta(R-1)t)\,f^f(z,t) + \int_0^t I_1\big(2\beta\sqrt{(R-1)\tau(t-\tau)}\big) \\ &\quad \sqrt{\frac{\beta^2(R-1)\tau}{t-\tau}}\,\exp\big(-\beta(t-\tau)\big)\exp\big(-\beta(R-1)\tau\big)\,f^f(z,\tau)\,d\tau\,, \end{aligned} \tag{A.47}$$

where

$$f^f(z,t) = \frac{z}{2\sqrt{\pi Dt^3}}\exp\Big(-\frac{(z-Vt)^2}{4Dt}\Big)\,. \tag{A.48}$$

A.2 The Fourier Transform

The Fourier transform $\tilde{f}(\lambda)$ of a function $f(z)$; $z \in [-\infty, +\infty]$ is defined by[2]

$$\tilde{f}(\lambda) \equiv \mathcal{F}(f(z)) := \int_{-\infty}^{\infty} f(z) \exp(-i\lambda z)\, dz \ , \tag{A.49}$$

if the integral in (A.49) exists.

The corresponding inverse transform is given by

$$f(z) = \mathcal{F}^{-1}(\tilde{f}(\lambda)) := \frac{1}{2\pi} \int_{-\infty}^{\infty} \tilde{f}(\lambda) \exp(i\lambda z)\, d\lambda \ , \tag{A.50}$$

where λ is the wave length conjugate to z, and $i := \sqrt{-1}$. In contrast to the inverse integral (A.31) of the Laplace transform, the inverse Fourier transform (A.50) is an ordinary definite integral. Therefore, any tabulation of definite integrals can be used to invert the Fourier transform.

One may also define a three-dimensional Fourier transform of the vector $\mathbf{x} = (x_1, x_2, x_3)$ in terms of the wave vector $\boldsymbol{\lambda} = (\lambda_1, \lambda_2, \lambda_3)$:

$$\tilde{f}(\boldsymbol{\lambda}) := \int_{-\infty}^{\infty} \int_{-\infty}^{\infty} \int_{-\infty}^{\infty} f(\mathbf{x}) \exp(-i\boldsymbol{\lambda} \cdot \mathbf{x})\, dx_1 dx_2 dx_3 \ , \tag{A.51}$$

$$f(\mathbf{x}) := \frac{1}{(2\pi)^3} \int_{-\infty}^{\infty} \int_{-\infty}^{\infty} \int_{-\infty}^{\infty} \tilde{f}(\boldsymbol{\lambda}) \exp(i\boldsymbol{\lambda} \cdot \mathbf{x})\, d\lambda_1 d\lambda_2 d\lambda_3 \ , \tag{A.52}$$

where $\boldsymbol{\lambda} \cdot \mathbf{x} := \lambda_1 x_1 + \lambda_2 x_2 + \lambda_3 x_3$ is the inner vector product of $\boldsymbol{\lambda}$ and $\mathbf{x}$ (Arfken, 1985).

Example A.7 FOURIER TRANSFORM OF THE EXPONENTIAL COVARIANCE FUNCTION

The three dimensional, isotropic, exponential covariance function (7.42) introduced in Chapter 7 may be written as

$$\mathrm{Cov}(r) = \sigma_Y^2 \exp(-r/L_Y) \ , \tag{A.53}$$

where $r = \sqrt{x^2 + y^2 + z^2}$ is the separation between locations of the random variable $Y = \ln(K_s)$. The Fourier transform (A.51) of (A.53) is most easily evaluated by changing to spherical coordinates

$$\mathcal{F}\big(\mathrm{Cov}(r)\big) = \sigma_Y^2 \int_0^{\infty} \int_0^{\pi} \int_0^{2\pi} \exp\Big(-\frac{r}{L_Y} + i\lambda r \cos(\theta)\Big) r^2 \sin(\theta)\, dr d\theta d\phi \ . \tag{A.54}$$

The integrations over θ and ϕ may be carried out easily, reducing (A.54) to

$$\mathcal{F}\big(\mathrm{Cov}(r)\big) = \frac{2\pi\sigma_Y^2}{i\lambda} \int_0^{\infty} \left[\exp\left[r\left(i\lambda - \frac{1}{L_Y}\right)\right] - \exp\left[-r\left(i\lambda + \frac{1}{L_Y}\right)\right]\right] r\, dr \ , \tag{A.55}$$

and finally, using the general integral form

$$\int_0^{\infty} x^n \exp(-ax)\, dx = \frac{n!}{a^{n+1}} \ , \tag{A.56}$$

[2]There are alternate forms of this definition. We choose to define it in this way to utilize the same structure as the Laplace transform above.

we reduce (A.55) to

$$\mathcal{F}\big(\operatorname{Cov}(r)\big) = \frac{8\pi\sigma_Y^2 L_Y^3}{\big(1+(\lambda L_Y)^2\big)^2} \; . \tag{A.57}$$

The Fourier transform of the autocorrelation function appears in the formulation of the macrodispersion coefficients on stochastic continuum theory (Gelhar and Axness, 1983; Dagan, 1984; 1987). (The expression in (A.57) differs by a constant factor from Dagan's 1984 formula because of the difference in the definition of the Fourier transform.)

Example A.8 INVERSION OF THE FOURIER TRANSFORM OF THE RESIDENT PDF OF THE CDE IN INFINITE SOIL

In Example 3.6, it was shown that the Fourier transform of the resident pdf of the CDE in infinite soil is equal to (3.56)

$$\widehat{C_t^r}(\lambda;t) = \exp(-i\lambda Vt - \lambda^2 Dt) \; . \tag{A.58}$$

Therefore, the inverse transform (A.50) may be written as

$$C_t^r(z,t) = \frac{1}{2\pi}\int_{-\infty}^{\infty} \exp(-\lambda^2 Dt + i\lambda(z-Vt))\, d\lambda \; . \tag{A.59}$$

The expression in the exponential of (A.59) may be rewritten as

$$-\lambda^2 Dt + i\lambda(z-Vt) = -Dt\Big(\lambda - \frac{i(z-Vt)}{2Dt}\Big)^2 - \frac{(z-Vt)^2}{4Dt} \; . \tag{A.60}$$

Therefore, after inserting (A.60) into (A.59), and letting $y := \sqrt{Dt}(\lambda - i(z-Vt)/2Dt)$, we may write (A.59) as

$$\begin{aligned} C_t^r(z,t) &= \frac{1}{2\pi\sqrt{Dt}} \exp\Big(-\frac{(z-Vt)^2}{4Dt}\Big)\int_{-\infty}^{\infty} \exp(-y^2)\, dy \\ &= \frac{1}{2\sqrt{\pi Dt}} \exp\Big(-\frac{(z-Vt)^2}{4Dt}\Big) \; . \end{aligned} \tag{A.61}$$

Appendix B

Useful Fortran Computer Programs

B.1 Numerical Inversion of Laplace Transforms

The following program is written in Fortran 77, and uses complex variables. Function *FS* is the Laplace transform of the desired solution, which is inverted for each value of t furnished from the main program by calling the subroutine Talbot. This subroutine returns the value *FT* of the function corresponding to the input value of t. The integer N is the number of terms requested for the series expansion in the subroutine. The value 63 is arbitrary, but seems to be optimal for a variety of problems we encountered.

In the numerical example provided, the Laplace transform of the CDE flux pdf (2.50) is inverted numerically, and compared to the analytic solution (2.51). Appreciation is expressed to Garrison Sposito and J.A. Barker for the development of this program.

```
      PROGRAM FLUXPDF

      IMPLICIT REAL*8(A-H,O-Z)
      REAL J
      COMMON /ARR/Z,V,D
      OPEN(UNIT=2,FILE="CDE.OUT",STATUS="NEW")
      V=2.
      Z=30.
      D=3.
      N=63
      PI=3.141592653589
      DO 122 J=1,25.
        T=DFLOAT(J)
        Y=(Z-V*T)/DSQRT(4.*D*T)
        IF(DABS(Y).GT.10.)THEN
            F1=0.
```

```
          ELSE
              F1=DEXP(-Y*Y)*Z/SQRT(4.*PI*D*T**3)
          END IF
          CALL TALBOT(FT,T,N)
          WRITE(*,125)T,FT,F1
          WRITE(2,125)T,FT,F1
122     CONTINUE
125     FORMAT(3(F10.7,2X))
        PAUSE
        END
C----------------------------------------------------------------------
        SUBROUTINE TALBOT(FT,T,N)
        IMPLICIT REAL*8(A-H,O-Z)
        REAL*8 NU,LAMDA
        COMPLEX S(80),DS(80),ZZ,SUM,B,B1,B2,FS,S9
        DATA Z/0.0D0/,PI/3.1415926535897932D0/,NU/1.0D0/,TAU/6.0D0/
        PIBYN=PI/DFLOAT(N)
        AA=1.
        BB=0.5
        CC=2.
        ZZ=CMPLX(Z,Z)
        LAMDA=TAU/T
        NMI=N-1
        DO 10 K=1,NMI
          U=DFLOAT(K)
          THETA=U*PIBYN
          ALPHA=THETA*DCOS(THETA)/DSIN(THETA)
          S(K)=CMPLX(ALPHA,THETA)
          DS(K)=CMPLX(NU,THETA+ALPHA*(ALPHA-AA)/THETA)*BB
   10 CONTINUE
        PSI=TAU*PIBYN
        CP=2.*DCOS(PSI)
        SP=DSIN(PSI)
        B=ZZ
        B1=B
        DO 3 KA=1,NMI
          K=N-KA
          V2=DEXP(DMAX1(TAU*DREAL(S(K)),-1.8D+02))
          B2=B1
          B1=B
          B=CP*B1-B2+V2*DS(K)*FS(LAMDA*S(K),IND)
    3 CONTINUE
        SUM=DEXP(TAU)*FS(LAMDA+ZZ,IND)*BB+CP*B-(B1-B*CMPLX(Z,SP))*CC
        FT=LAMDA*REAL(SUM)/DFLOAT(N)
        RETURN
        END
C----------------------------------------------------------------------
```

```
      FUNCTION FS(S9,IND)
C-----LAPLACE TRANSFORMATION OF THE CDE FLUX PDF
      IMPLICIT REAL*8(A-H,O-Z)
      COMMON /ARR/Z9,V9,D9
      COMPLEX FS,S9,XI
      XI=CSQRT(1.+4.*S9*D9/V9**2)
      FS=CEXP(Z9*V9/2./D9*(1.-XI))
      RETURN
      END
```

Time	Inversion	Analytic
1.0000000	.0000000	.0000000
2.0000000	.0000000	.0000000
3.0000000	.0000001	.0000001
4.0000000	.0000255	.0000255
5.0000000	.0005562	.0005562
6.0000000	.0036932	.0036932
7.0000000	.0125240	.0125240
8.0000000	.0280308	.0280308
9.0000000	.0477016	.0477016
10.0000000	.0671496	.0671496
11.0000000	.0824707	.0824707
12.0000000	.0915398	.0915399
13.0000000	.0940800	.0940801
14.0000000	.0910801	.0910801
15.0000000	.0841044	.0841044
16.0000000	.0747701	.0747701
17.0000000	.0644496	.0644496
18.0000000	.0541583	.0541582
19.0000000	.0445573	.0445572
20.0000000	.0360126	.0360126
21.0000000	.0286720	.0286720
22.0000000	.0225367	.0225367
23.0000000	.0175202	.0175202
24.0000000	.0134915	.0134914
25.0000000	.0103036	.0103035

B.2 Error Function Evaluation

This subroutine provides values for $\mathrm{erf}(X)$ or $\exp(A)\,\mathrm{erfc}(X)$. The subroutine $\mathrm{erf}(X)$ is called from a main program by `CALL ERF(x,B)`, and the value returned for B is $\mathrm{erf}(X)$. The complementary error function $\mathrm{erfc}(X)$ frequently appears in an expression in which it is multiplied by an exponential with a positive argument. When the exponential becomes large, it must be evaluated asymptotically with erfc(X). This is done automatically when necessary in the program. The function exp(A) erfc(X) is called from the main program by `CALL ERFC(X,A,B)`. The value of B is exp(A) erfc(X). To evaluate erfc(X) alone, the command `CALL ERFC(X,0,B)` may be used, or the definition erfc(x) := 1 − erf(x). The numerical approximations for the error function are taken from Abramowitz and Stegun, (1970).

```
      SUBROUTINE ERF(ARG,B)
      IMPLICIT REAL*8(A-H,O-Z)
      STORE=0.
C-----THIS IS THE ERROR FUNCTION SUBROUTINE. FEED IN A AND RETURN B=ERF(A).
      IF (ARG.LT.0.) THEN
        STORE=ARG
        ARG=-ARG
      ENDIF
      B=1./(1.+.3275911*ARG)
      BTEM=.254829592*B-.284496736*B*B+1.421413741*B*B*B
     &     -1.453152027*B*B*B*B
      IF (DABS(ARG).GT.170.) THEN
        B=1.
        GOTO 1
      ENDIF
      B=1.-(BTEM+1.061405429*B**5.)*DEXP(-ARG*ARG)
    1 IF(STORE.LT.0.) THEN
        B=-B
        ARG=-ARG
      ENDIF
      RETURN
      END
C-----------------------------------------------------------------------
      SUBROUTINE ERFC(ARG,EXPARG,B)
      IMPLICIT REAL*8(A-H,O-Z)
      IF(ARG.LT.0.) THEN
        CALL ERF(ARG,B)
        B=DEXP(EXPARG)*(1.-B)
        GO TO 1
      ENDIF
      IF(ARG.GT.3.5) THEN
        CALL ERFC2(ARG,EXPARG,B)
        GO TO 1
      ENDIF
      B=1./(1.+.3275911*ARG)
      BTEM=.254829592*B-.284496736*B*B+1.42143741*B*B*B
     &     -1.453152027*B*B*B*B
      IF(DABS(-ARG*ARG+EXPARG).GT.170.) THEN
        B=0.0
```

```
        GO TO 1
      ENDIF
      B=(BTEM+1.061405429*B**5.)*DEXP(-ARG*ARG+EXPARG)
    1 RETURN
      END
C-----------------------------------------------------------------------
      SUBROUTINE ERFC2(ARG,EXPARG,B)
      IMPLICIT REAL*8(A-H,O-Z)
      PI=3.141592654
      IF(DABS(-ARG*ARG+EXPARG).GT.170.) THEN
        B=0.0
        RETURN
      ENDIF
      B=DEXP(-ARG*ARG+EXPARG)/DSQRT(PI)*(1./ARG-1./(2.*ARG*ARG*ARG)
     &  +3./(4.*ARG**5.))
      RETURN
      END
```

Appendix C

Table of Laplace Transforms

This table is compiled from entries found in Van Genuchten and Alves (1982), Abramowitz and Stegun (1970), and Walker (1987). The first reference provided the source for the system of abbreviations used below.

$$\mathcal{A} := \frac{1}{\sqrt{\pi t}} \exp\Big(-\frac{x^2}{4t}\Big)$$

$$\mathcal{B} := \operatorname{erfc}\Big(\frac{x}{2\sqrt{t}}\Big)$$

$$\mathcal{C} := \exp(a^2 t - ax) \operatorname{erfc}\Big(\frac{x}{2\sqrt{t}} - a\sqrt{t}\Big)$$

$$\mathcal{D} := \exp(a^2 t + ax) \operatorname{erfc}\Big(\frac{x}{2\sqrt{t}} + a\sqrt{t}\Big)$$

I_N is the modified Bessel function of integer order N; a and b are constants.

$f(t)$	$\hat{f}(s) := \int_0^\infty f(t)\, exp(-st)\, dt$	
$\delta(t)$	1	(C.1)
1	$\frac{1}{s}$	(C.2)
t^N	$\frac{N!}{s^{N+1}}$	(C.3)
$\exp(-at)$	$\frac{1}{s+a}$	(C.4)
$\frac{\sin(at)}{a}$	$\frac{1}{s^2+a^2}$	(C.5)

$$\cos(at) \qquad \frac{s}{s^2+a^2} \tag{C.6}$$

$$\frac{x}{2t}\mathcal{A} \qquad \exp(-x\sqrt{s}) \tag{C.7}$$

$$\mathcal{A} \qquad \frac{\exp(-x\sqrt{s})}{\sqrt{s}} \tag{C.8}$$

$$\mathcal{B} \qquad \frac{\exp(-x\sqrt{s})}{s} \tag{C.9}$$

$$2t\mathcal{A} - x\mathcal{B} \qquad \frac{\exp(-x\sqrt{s})}{s\sqrt{s}} \tag{C.10}$$

$$\frac{\mathcal{C}+\mathcal{D}}{2} \qquad \frac{\exp(-x\sqrt{s})}{s-a^2} \tag{C.11}$$

$$\frac{\mathcal{C}-\mathcal{D}}{2a} \qquad \frac{\exp(-x\sqrt{s})}{\sqrt{s}(s-a^2)} \tag{C.12}$$

$$\mathcal{A} - a\mathcal{D} \qquad \frac{\exp(-x\sqrt{s})}{\sqrt{s}+a} \tag{C.13}$$

$$\mathcal{D} \qquad \frac{\exp(-x\sqrt{s})}{\sqrt{s}(\sqrt{s}+a)} \tag{C.14}$$

$$\frac{\mathcal{B}-\mathcal{D}}{a} \qquad \frac{\exp(-x\sqrt{s})}{s(\sqrt{s}+a)} \tag{C.15}$$

$$t\mathcal{A} + \frac{\mathcal{C}}{4a} - \frac{\mathcal{D}}{4a}(1+2ax+4a^2t) \qquad \frac{\exp(-x\sqrt{s})}{(s-a^2)(\sqrt{s}+a)} \tag{C.16}$$

$$-\frac{t\mathcal{A}}{a} + \frac{\mathcal{C}}{4a^2} + \frac{\mathcal{D}}{4a^2}(-1+2ax+4a^2t) \qquad \frac{\exp(-x\sqrt{s})}{\sqrt{s}(s-a^2)(\sqrt{s}+a)} \tag{C.17}$$

$$\frac{t\mathcal{A}}{a^2} - \frac{\mathcal{B}}{a^3} + \frac{\mathcal{C}}{4a^3} + \frac{\mathcal{D}}{4a^3}(3-2ax-4a^2t) \qquad \frac{\exp(-x\sqrt{s})}{s(s-a^2)(\sqrt{s}+a)} \tag{C.18}$$

$$(1+ax+2a^2t)\mathcal{D} - 2at\mathcal{A} \qquad \frac{\exp(-x\sqrt{s})}{(\sqrt{s}+a)^2} \tag{C.19}$$

$$2t\mathcal{A} - (x+2at)\mathcal{D} \qquad \frac{\exp(-x\sqrt{s})}{\sqrt{s}(\sqrt{s}+a)^2} \tag{C.20}$$

$$\frac{\mathcal{D}}{a^2}(-1+ax+2a^2t) + \frac{\mathcal{B}}{a^2} - \frac{2t}{a}\mathcal{A} \qquad \frac{\exp(-x\sqrt{s})}{s(\sqrt{s}+a)^2} \tag{C.21}$$

$$\frac{\mathcal{C}}{8a^2} - \frac{t\mathcal{A}}{2a}(1+ax+2a^2t) + \frac{\mathcal{D}}{8a^2}(-1+2ax+8a^2t+2a^2(x+2at)^2) \qquad \frac{\exp(-x\sqrt{s})}{(s-a^2)(\sqrt{s}+a)^2} \tag{C.22}$$

$$I_0(2\sqrt{at}) \qquad \frac{1}{s}\exp\left(\frac{a}{s}\right) \tag{C.23}$$

$$I_0(2\sqrt{at})\ \exp(-bt) \qquad \frac{1}{s+b}\exp\left(\frac{a}{s+b}\right) \tag{C.24}$$

$$\delta(t) + I_1(2\sqrt{at})\sqrt{\frac{a}{t}}\exp(-bt) \qquad \exp\left(\frac{a}{s+b}\right) \tag{C.25}$$

$$1 - \int_0^a \exp(-y - bt)\ I_0(2\sqrt{ayt})\ dy =: J(a, bt) \qquad \frac{1}{s}\ \exp\left(-\frac{as}{s+b}\right) \tag{C.26}$$

(Goldstein's J-function)

Appendix D

Solutions to Problems

D.1 Problems in Chapter 1

Solution 1.1

If the system is perfectly mixed at all times, then any solute mass added to the saline inlet will instantly cause the concentration $C_V(t)$ inside the vessel of water volume V to assume a new and uniform value. Thus, we can write a mass balance equation for the solute in the mixing tank, which in words is:

rate of input of mass into the tank −
rate of removal of mass from the tank =
rate of increase of mass stored in the tank

Substituting the appropriate parameters from the problem, we have

rate of input of mass into tank $= C_s(t)Q_s$
rate of removal of mass from the tank $= C_V(t)Q_o$
rate of increase of mass stored in the tank $= V\, dC_V(t)/dt$

Therefore, the differential equation describing the perfectly mixed system is

$$V\frac{dC_V(t)}{dt} = Q_s C_s(t) - Q_o C_V(t) \; . \tag{D.1}$$

To solve for the travel time pdf $f(t)$ we must solve (D.1) subject to the following initial and boundary conditions

$$C_V(0) = 0 \quad , \quad C_s(t) = \frac{Q_o}{Q_s}\delta(t) \; . \tag{D.2}$$

The delta function (see (2.8)) describing the saline input concentration is weighted by the flow ratio Q_o/Q_s so that the integral of the outflow concentration over time will be unity.[1]

[1] The reader may notice that the input flux concentration is $C^f = Q_s C_s/Q_o = \delta(t)$.

The easiest way to solve (D.1) subject to (D.2) is by Laplace transforms. (Readers not familiar with this approach are encouraged to read Appendix A.) The Laplace transform of (D.1) and (D.2) is

$$sV\widehat{C}_V = Q_o - Q_o\widehat{C}_V \ , \tag{D.3}$$

where

$$\widehat{C}_V = \int_0^\infty C_V(t)\, \exp(-st)\, dt \tag{D.4}$$

is the Laplace tranform of $C_V(t)$ and s is a parameter. The solution of (D.3) is

$$\widehat{C}_V = \frac{Q_o}{sV + Q_o} \ . \tag{D.5}$$

This solution may be inverted using the Table of Laplace Transforms ((C.4) of Appendix C), which produces

$$f(t) = C_V(t) = \frac{Q_o}{V}\, \exp\left(\frac{Q_o t}{V}\right) = \frac{\exp(-t/\tau)}{\tau} \ . \tag{D.6}$$

This is the impulse response function or travel time pdf for the perfectly mixed tank, which is characterized by the mixing time $\tau = V/Q_o$.

Solution 1.2

When $C_s(t)$ is time dependent, the outflow concentration is given by (1.4) and (D.6)

$$C_o(t) = C_V(t) = \int_0^t C_s(t - t')\frac{\exp(-t'/\tau)}{\tau}\, dt' \ . \tag{D.7}$$

To solve for the outflow concentration when $Q_s(t)$ is time-dependent, we return to the differential equation (D.1), now in the form

$$V\frac{dC_V(t)}{dt} = Q_s(t)C_s(t) - Q_oC_V(t) \ . \tag{D.8}$$

We again take the Laplace transform of this equation and solve for C_V, this time obtaining

$$\widehat{C}_V = \frac{\widehat{G}_s}{sV_V + Q_o} \ , \tag{D.9}$$

where $\widehat{G}_s$ is the Laplace transform of the solute mass input flow rate $C_s(t)Q_s(t)$. Since $\widehat{C}_V$ in (D.9) is the product of two Laplace transforms, the inverse transform is obtained by applying the convolution integral (A.37)

$$Q_oC_o(t) = \int_0^t Q_s(t - t')\, C_s(t - t')\frac{\exp(-t'/\tau)}{\tau}\, dt' \ . \tag{D.10}$$

Equation (D.10) may be written in the form

$$q_{out}(t) = \int_0^t q_{in}(t - t')\frac{\exp(-t'/\tau)}{\tau}\, dt' \ , \tag{D.11}$$

where $q(t) = Q(t)C(t)$ is the solute mass flow rate. Thus, in the case of perfect mixing, the characteristic mixing time of the system is independent of the saline water flow rate, as long as the total flow rate is a constant.

Solution 1.3

The only step required to transform the integral is a change of variable to $\tau = t - t'$. Thus,

$$\int_0^t C_s(t-t')\, f(t')\, dt' = -\int_t^0 C_s(\tau)\, f(t-\tau)\, d\tau = \int_0^t C_s(\tau)\, f(t-\tau)\, d\tau \; . \tag{D.12}$$

The integral is the input concentration at time τ multiplied by the probability that it has a travel time $t - \tau$, summed over all possible values of τ from 0 to t.

Solution 1.4

Any nonlinearity in the system would violate the assumption that the concentrations will superimpose on top of each other at the outlet end. Also, the impulse response function would no longer contain all of the information about the travel time distribution, since the latter would depend on the number of solute molecules in the system.

D.2 Problems in Chapter 2

Solution 2.1

Since the transform variable s only appears in the expression (2.48) for ξ, we may use the shifting theorem (A.33) to evaluate the inverse transform. First, we write ξ in the form

$$\xi = \sqrt{1 + \frac{4sD}{V^2}} = \frac{2\sqrt{D}}{V}\sqrt{s + \frac{V^2}{4D}} \; . \tag{D.13}$$

Then, using (A.33) with $a = 1$ and $b = V^2/4D$, we obtain

$$\mathcal{L}^{-1}\left(\exp\left(\frac{Vz}{2D}(1-\xi)\right)\right) = \exp\left(\frac{Vz}{2D} - \frac{V^2 t}{4D}\right)\mathcal{L}^{-1}\left(\exp(-x\sqrt{s})\right) \; , \tag{D.14}$$

where

$$x = \frac{z}{\sqrt{D}} \; . \tag{D.15}$$

The inverse Laplace transform operation $\mathcal{L}^{-1}\left(\exp(-x\sqrt{s})\right)$ is given by (C.7),

$$\mathcal{L}^{-1}\left(\exp(-x\sqrt{s})\right) = \frac{z}{2\sqrt{\pi D t^3}} \exp\left(-\frac{z^2}{4Dt}\right) \; . \tag{D.16}$$

Insertion of (D.16) into (D.14) yields (2.51).

Solution 2.2

This is most easily shown after Laplace transformation. The Laplace transform of the stochastic-convective Fickian pdf may be written by combining (2.50) and (2.69), as

$$\widehat{f}^f(z;s) = \exp\Big(\frac{V\ell}{2D_\ell}(1-\xi')\Big) \quad , \quad \xi' = \sqrt{1+\frac{4sD_\ell z}{\ell V^2}} \ . \tag{D.17}$$

If $f^f(z,t)$ in (2.51) with (2.69) satisfies the CDE (2.37), then (D.17) must satisfy the Laplace-transformed CDE (2.42). By direct differentiation, we obtain, using the chain rule

$$\frac{d\widehat{f}}{dz} = \frac{d\xi'}{dz}\frac{d\widehat{f}}{d\xi'} = \frac{2sD_\ell}{\ell V^2\xi'}\Big(-\frac{V\ell}{2D_\ell}\Big)\widehat{f} = -\frac{s}{V\xi'}\widehat{f} \tag{D.18}$$

and

$$\frac{d^2\widehat{f}}{dz^2} = \Big[\frac{2sD_\ell}{\ell V^2\xi'}\frac{s}{V\xi'^2} + \Big(\frac{s}{V\xi'}\Big)^2\Big]\widehat{f} = \Big(1+\frac{2D_\ell}{\ell V\xi'}\Big)\Big(\frac{s}{V\xi'}\Big)^2\widehat{f} \ . \tag{D.19}$$

Inserting these expressions into the Laplace-transformed CDE, we obtain

$$D(z)\frac{d^2\widehat{f}}{dz^2} - V\frac{d\widehat{f}}{dz} - s\widehat{f} = \Big[D(z)\Big(1+\frac{2D_\ell}{\ell V\xi'}\Big)\Big(\frac{s}{V\xi'}\Big)^2 + \frac{s}{\xi'} - s\Big]\widehat{f} \neq 0 \ , \tag{D.20}$$

where $D(z) = zD_\ell/\ell$.

Solution 2.3

Inspection of (2.49) reveals that in general

$$\int_0^\infty C^f(0,t)\,\exp(-st)\,dt = \widehat{C}^f(0) = A \ . \tag{D.21}$$

Thus, A is the Laplace transform of the inlet flux concentration. The transforms of (2.73)–(2.75) are straightforward, with the result

$$A = \int_0^\infty \delta(t)\,\exp(-st)\,dt = 1 \ , \tag{D.22}$$

$$A = \int_0^\infty H(t)\,\exp(-st)\,dt = \frac{1}{s} \ , \tag{D.23}$$

$$\begin{aligned} A &= \int_0^\infty (H(t) - H(t-\Delta t))\,\exp(-st)\,dt \\ &= \int_0^{\Delta t} \exp(-st)\,dt = \frac{1}{s}(1-\exp(-s\Delta t)) \ . \end{aligned} \tag{D.24}$$

Solution 2.4

If we insert (2.76) into (2.77) and make the change of variable substitution

$$y = \frac{\ln(t) - \mu}{\sigma} \tag{D.25}$$

then the various parts of (2.77) are changed as follows

$$\frac{dt}{\sigma t} = dy \ , \qquad t = \exp(\mu + \sigma y) \ , \qquad \int_0^\infty \frac{dt}{\sigma t} \longrightarrow \int_{-\infty}^\infty dy \ . \tag{D.26}$$

Thus, with these substitutions, (2.77) becomes

$$\begin{aligned} \int_0^\infty t^N f(t)\, dt &= \int_{-\infty}^\infty \frac{\exp(N\mu + N\sigma y - y^2/2)}{\sqrt{2\pi}}\, dy \\ &= \exp(N\mu + N^2\sigma^2/2) \int_{-\infty}^\infty \exp\Big(-\frac{(y-\sigma)^2}{2}\Big)\, dy \ . \end{aligned} \tag{D.27}$$

Then, letting $z = y - \sigma$, and recognizing that

$$\int_{-\infty}^\infty \frac{\exp(-z^2/2)}{\sqrt{2\pi}}\, dz = 1 \tag{D.28}$$

we obtain (2.77).

Solution 2.5

By assumption, $f_1(t)$ and $f_2(t)$ are travel time pdfs and therefore have unit area when plotted against time. Each of them can be obtained from the same fundamental curve $f^*(I)$ where $I = J_w t$. However, since the fundamental curve is normalized when plotted against I, we must rescale the y axis of the f^* curve to obtain f_1 or f_2.

Therefore,

$$f_1(t) = J_1\, f^*(J_1 t) \tag{D.29}$$

$$f_2(t) = J_2\, f^*(J_2 t) \ . \tag{D.30}$$

We then rewrite (D.30) by multiplying by J_1/J_1 in several places, and then use (D.29) for $f_1(t)$. Thus,

$$f_2(t) = \frac{J_2}{J_1} J_1\, f^*\Big(J_1 \frac{J_2 t}{J_1}\Big) = \frac{J_2}{J_1} f_1\Big(\frac{J_2 t}{J_1}\Big) \ . \tag{D.31}$$

Solution 2.6

According to Problem 2.3, the Laplace transform of the solution to the CDE subject to the condition in (2.78) is

$$\widehat{C}^f(z;s) = \frac{C_0}{s} \exp\left(\frac{Vz}{2D}(1-\xi)\right) . \tag{D.32}$$

If we again use the shifting theorem (A.33) as we did in Problem 2.1, we obtain

$$\mathcal{L}^{-1}\left[\frac{C_0}{s} \exp\left(\frac{Vz}{2D}(1-\xi)\right)\right] = C_0 \exp\left(\frac{Vz}{2D} - \frac{V^2t}{4D}\right) \mathcal{L}^{-1}\left[\frac{\exp(-x\sqrt{s})}{s - V^2/4D}\right] \quad , \quad x := \frac{z}{\sqrt{D}} . \tag{D.33}$$

The remaining transform is of the form given in (C.11), with $a = V/2\sqrt{D}$. After some simplification, we obtain

$$C^f(z,t) = \frac{C_0}{2}\left[\operatorname{erfc}\left(\frac{z-Vt}{\sqrt{4Dt}}\right) + \exp\left(\frac{Vz}{D}\right) \operatorname{erfc}\left(\frac{z+Vt}{\sqrt{4Dt}}\right)\right] . \tag{D.34}$$

This solution was first presented by Lapidus and Amundson (1952).

Solution 2.7

The CDE (2.37) with boundary conditions appropriate for the flux pdf solution $f^f(z, I)$ may be written as

$$\theta\frac{\partial C^f}{\partial t} + J_w\frac{\partial C^f}{\partial z} - \theta D\frac{\partial^2 C^f}{\partial z^2} = 0 ; \tag{D.35}$$

$$C^f(z,0) = 0 \quad , \quad C^f(\infty,t) = 0 \quad , \quad C^f(0,t) = \frac{\delta(t)}{J_w} , \tag{D.36}$$

where the factor of $1/J_w$ in the inlet flux concentration in (D.36) is required so that the area under the curve of C^f versus I will be unity. If we make the substitution $I = J_w t$ in (D.35) and (D.36), we obtain

$$\theta\frac{\partial C^f}{\partial I} + \frac{\partial C^f}{\partial z} - \lambda\frac{\partial^2 C^f}{\partial z^2} = 0 ; \tag{D.37}$$

$$C^f(z,0) = 0 \quad , \quad C^f(\infty,I) = 0 \quad , \quad C^f(0,I) = \delta(I) , \tag{D.38}$$

where $\lambda := \theta D/J_w = D/V$. The solution to these equations is of the form $C^f(z, I; \theta, \lambda)$, which is invariant as a function of I for fixed z if θ and λ are constant.

Solution 2.8

Using the general definite integral (Abramowitz and Stegun, 1970)

$$\int_0^\infty x^a \exp(-bx)\, dx = \frac{a!}{b^{1+a}} \frac{\Gamma(1+a)}{b^{1+a}} , \tag{D.39}$$

where $\Gamma(x) := (x-1)!$ is the gamma function, the Nth travel time moment of (2.61) may be evaluated as

$$\mathrm{E}(t^N) = \frac{\beta^{1+\alpha}}{\alpha!} \int_0^\infty t^{\alpha+N} \exp(-\beta t)\, dt = \frac{\beta^{1+\alpha}}{\alpha!} \frac{(\alpha+N)!}{\beta^{1+\alpha+N}} = \frac{(\alpha+N)!}{\beta^N \alpha!} . \tag{D.40}$$

Thus, the mean, second moment, and variance of the gamma pdf (2.61) are

$$\mathrm{E}(t) = \frac{1+\alpha}{\beta} \quad , \quad \mathrm{E}(t^2) = \frac{(1+\alpha)(2+\alpha)}{\beta^2} \quad , \quad \mathrm{Var}(t) = \frac{1+\alpha}{\beta^2} \; . \tag{D.41}$$

Solution 2.9

Since s is a constant in the Laplace transform integral (A.1) over t, we may exchange the order of integration and differentiation. Thus,

$$\begin{aligned} \mathrm{E}(t^N) &= (-1)^N \frac{d^N \widehat{f}(s)}{ds^N} = (-1)^N \frac{d^N}{ds^N} \int_0^\infty f(t)\, \exp(-st)\, dt \\ &= (-1)^N \int_0^\infty f(t)\, \frac{d^N \exp(-st)}{ds^N}\, dt \\ &= (-1)^N \int_0^\infty f(t)(-t)^N \exp(-st)\, dt \\ &\overset{s=0}{\longrightarrow} \int_0^\infty t^N f(t)\, dt \; . \end{aligned} \tag{D.42}$$

Solution 2.10

If we begin with (2.59)

$$f^f(\ell, t) = \frac{\ell}{2\sqrt{\pi D t^3}} \exp\Big(-\frac{(\ell - Vt)^2}{4Dt}\Big) \; , \tag{D.43}$$

then

$$\begin{aligned} \frac{\ell}{z} f^f\Big(\ell, \frac{t\ell}{z}\Big) &= \frac{\ell^2}{2z\sqrt{\pi D \ell^3 t^3 / z^3}} \exp\Big(-\frac{(\ell - Vt\ell/z)^2}{4Dt\ell/z}\Big) \\ &= \frac{z}{2\sqrt{\pi (Dz/\ell) t^3}} \exp\Big(-\frac{(z - Vt)^2}{4(Dz/\ell)t}\Big) \; , \end{aligned} \tag{D.44}$$

which is not the same as (2.51). Thus, (2.51) with constant D is not stochastic-convective. However, (2.51) with $D(z) = Dz/\ell$ is equal to (D.44), and therefore is stochastic-convective.

D.3 Problems in Chapter 3

Solution 3.1

We wish to solve the resident form of the CDE (2.36) subject to the following initial and boundary conditions

$$C_l^r(z,0) = 0 \quad , \quad C_l^r(\infty,t) = 0 \ , \tag{D.45}$$

$$\left(J_w C_l^r - \theta D \frac{\partial C_l^r}{\partial z}\right)\bigg|_{z=0} = J_w \delta(t) \ . \tag{D.46}$$

After Laplace transformation and application of conditions (D.45), the solution for the resident concentration may be written (see Example 2.2)

$$\widehat{C}_l^r(z;s) = A \exp\left(\frac{Vz}{2D}(1-\xi)\right) \quad ; \quad \xi = \sqrt{1 + \frac{4sD}{V^2}} \ , \tag{D.47}$$

where A is constant. The Laplace transform of (D.46) is

$$\left(J_w \widehat{C}_l^r - \theta D \frac{\partial \widehat{C}_l^r}{\partial z}\right)\bigg|_{z=0} = J_w \ . \tag{D.48}$$

If we insert (D.47) into (D.48), we obtain

$$\left(J_w - \theta D \frac{V}{2D}(1-\xi)\right) A = J_w \ . \tag{D.49}$$

Since $\theta V = J_w$, A is equal to

$$A = \frac{2}{1+\xi} \ . \tag{D.50}$$

Thus

$$C_l^r(z;s) = \frac{2}{1+\xi} \exp\left(\frac{Vz}{2D}(1-\xi)\right) \ . \tag{D.51}$$

Applying the shifting theorem (A.33) as in Problem 2.1, we obtain

$$\mathcal{L}^{-1}\left[\frac{2}{1+\xi} \exp\left(\frac{Vz}{2D}(1-\xi)\right)\right] = \frac{V}{\sqrt{D}} \exp\left(\frac{Vz}{2D} - \frac{V^2 t}{4D}\right) \mathcal{L}^{-1}\left[\frac{\exp(-x\sqrt{s})}{\sqrt{s} + V/2\sqrt{D}}\right] \ , \tag{D.52}$$

where

$$x := \frac{z}{\sqrt{D}} \ . \tag{D.53}$$

The second inverse transform is of the form (C.13). After some simplification the inverse transform (D.52) reduces to (3.12).

Solution 3.2

If we plug the stochastic-convective flux pdf (2.65) into the expression (3.37) for the resident pdf, we obtain

$$\begin{aligned} f^r(z,t) &= -\frac{\partial}{\partial z}\int_0^t f^f(z,t)\,dt = -\frac{\partial}{\partial z}\int_0^t \frac{\ell}{z} f^f\left(\ell,\frac{t\ell}{z}\right)dt \\ &= -\frac{\partial}{\partial z}\int_0^{t\ell/z} f^f(\ell,y)\,dy = \frac{t\ell}{z^2} f^f\left(\ell,\frac{t\ell}{z}\right) = \frac{t}{z} f^f(z,t)\ , \end{aligned} \tag{D.54}$$

where we have used Leibnitz's rule (Arfken, 1985)

$$\begin{aligned} \frac{\partial}{\partial x}\int_{b(x)}^{a(x)} g(x,y)\,dy &= \frac{da(x)}{dx} g(x,a(x)) - \frac{db(x)}{dx} g(x,b(x)) \\ &\quad + \int_{b(x)}^{a(x)} \frac{\partial g(x,y)}{\partial x}\,dy\ . \end{aligned} \tag{D.55}$$

Solution 3.3

The Laplace transform of the Nth depth moment (3.43) may be written as

$$\widehat{Z}_N = \frac{4DN!}{V^2(\xi^2-1)}\left(\frac{2D}{V(\xi-1)}\right)^N = \frac{N!}{s}\left(\frac{2D}{V(\xi-1)}\right)^N \tag{D.56}$$

using (3.42). Thus,

$$\widehat{Z}_0 = \frac{1}{s} \quad \xrightarrow{\mathcal{L}^{-1}} \quad Z_0 = 1\ , \tag{D.57}$$

showing that the pdf is normalized. The first moment is, using (A.33),

$$\widehat{Z}_1 = \frac{2D}{Vs(\xi-1)} \quad \xrightarrow{\mathcal{L}^{-1}} \quad Z_1(t) = \sqrt{D}\exp(-a^2t)\,\mathcal{L}^{-1}\left(\frac{1}{(s-a^2)(\sqrt{s}-a)}\right)\ , \tag{D.58}$$

where $a = V/2\sqrt{D}$. Thus, using the inversion formula (C.16) with $x = 0$, we obtain

$$Z_1(t) = \sqrt{\frac{Dt}{\pi}}\exp(-a^2t) + \frac{D}{2V}\left[(1+4a^2t)\,\mathrm{erfc}(-a\sqrt{t}) - \mathrm{erfc}(a\sqrt{t})\right]\ . \tag{D.59}$$

For large t this can be approximated by

$$\frac{V\sqrt{t}}{2\sqrt{D}} \gg 1 \quad \Rightarrow \quad Z_1(t) \approx \frac{D}{V} + Vt\ . \tag{D.60}$$

The second moment may be written as

$$\widehat{Z}_2 = \frac{2}{s}\left(\frac{2D}{V(\xi-1)}\right)^2 \quad \xrightarrow{\mathcal{L}^{-1}} \quad Z_2(t) = 2D\exp(-a^2t)\mathcal{L}^{-1}\left(\frac{1}{(s-a^2)(\sqrt{s}-a)^2}\right)\ , \tag{D.61}$$

which is equal to, using (C.22),

$$\begin{aligned} Z_2(t) &= \left(\frac{2D}{V} + Vt\right)\sqrt{\frac{Dt}{\pi}}\exp(-a^2t) \\ &\quad + \left(\frac{D}{V}\right)^2\left[\operatorname{erfc}(a\sqrt{t}) + (8a^4t^2 + 8a^2t - 1)\operatorname{erfc}(-a\sqrt{t})\right] . \end{aligned} \tag{D.62}$$

The approximation for large t is

$$\frac{V\sqrt{t}}{2\sqrt{D}} \gg 1 \qquad \Rightarrow \qquad Z_2(t) \approx V^2t^2 + 4Dt - 2\left(\frac{D}{V}\right)^2 . \tag{D.63}$$

Therefore, the asymptotic depth variance is

$$\frac{V\sqrt{t}}{2\sqrt{D}} \gg 1 \qquad \Rightarrow \qquad \operatorname{Var}(Z(t)) = Z_2(t) - Z_1^2(t) \approx 2Dt - 3\left(\frac{D}{V}\right)^2 . \tag{D.64}$$

Solution 3.4

The stochastic-convective model of the resident concentration in this case becomes

$$C_t^r(z,t) = \begin{cases} C_0 \int_0^z f^r(z',t)\,dz' & ,\ 0 < z < \ell \,; \\ C_0 \int_{z-\ell}^z f^r(z',t)\,dz' & ,\ z > \ell \,, \end{cases} \tag{D.65}$$

where the resident pdf of the CLT is given by (3.45)

$$f^r(z,t) = \frac{1}{\sqrt{2\pi}\sigma_\ell z}\exp\left(-\frac{\left(\ln(t\ell/z) - \mu_\ell\right)^2}{2\sigma_\ell^2}\right) , \tag{D.66}$$

where ℓ is the reference depth at which μ_ℓ and σ_ℓ were measured. If we insert (D.66) into (D.65) and define a new variable

$$y(z) = -\frac{\ln(t\ell/z) - \mu_\ell}{\sqrt{2}\sigma_\ell} = \frac{\ln(z/t\ell) + \mu_\ell}{\sqrt{2}\sigma_\ell} \quad , \quad dy = \frac{dz}{\sqrt{2}\,\sigma_\ell z} , \tag{D.67}$$

equation (D.65) becomes

$$C_t^r(z,t) = \begin{cases} \frac{C_0}{\sqrt{\pi}} \int_{-\infty}^{y(z)} \exp(-y'^2)\,dy' & ,\ 0 < z < \ell \,; \\ \frac{C_0}{\sqrt{\pi}} \int_{y(z-\ell)}^{y(z)} \exp(-y'^2)\,dy' & ,z > \ell \,. \end{cases} \tag{D.68}$$

Equation (D.68) may be written in terms of the error function (3.13). Therefore

$$C_t^r(z,t) = \begin{cases} \frac{C_0}{2}\left[1 + \operatorname{erf}\left(\frac{\ln(z/t\ell) + \mu_\ell}{\sqrt{2}\sigma_\ell}\right)\right] & ,\ 0 < z < \ell \,; \\ \frac{C_0}{2}\left[\operatorname{erf}\left(\frac{\ln(z/t\ell) + \mu_\ell}{\sqrt{2}\sigma_\ell}\right) - \operatorname{erf}\left(\frac{\ln((z-\ell)/t\ell) + \mu_\ell}{\sqrt{2}\sigma_\ell}\right)\right] & ,\ z > \ell \,. \end{cases} \tag{D.69}$$

Solution 3.5

By extension of the piston flow Example 3.1, we may write the flux pdf for the two region transport volume as

$$f^f(z,t) = \frac{\theta_1}{\theta}\delta\left(t - \frac{z}{V_1}\right) + \frac{\theta_2}{\theta}\delta\left(t - \frac{z}{V_2}\right) . \tag{D.70}$$

Thus, the first travel time moment is

$$\mathrm{E}_z(t) = \int_0^\infty t' f^f(z,t')\, dt' = \frac{\theta_1}{\theta}\frac{z}{V_1} + \frac{\theta_2}{\theta}\frac{z}{V_2} \tag{D.71}$$

and the mean flux velocity V^f is obtained from (2.57)

$$\frac{1}{V^f} = \frac{\theta_1}{\theta}\frac{1}{V_1} + \frac{\theta_2}{\theta}\frac{1}{V_2} . \tag{D.72}$$

Equation (D.72) may be reformulated in terms of the water flux $J_w = \theta_1 V_1 = \theta_2 V_2$. Then, V^f may be written as

$$V^f = \frac{\theta J_w}{\theta_1^2 + \theta_2^2} . \tag{D.73}$$

The resident pdf obtained from the flux pdf using (3.37) is

$$\begin{aligned} f^r(z,t) &= -\frac{\partial}{\partial z}\left[\frac{\theta_1}{\theta}H\left(t - \frac{z}{V_1}\right) + \frac{\theta_2}{\theta}H\left(t - \frac{z}{V_2}\right)\right] \\ &= \frac{\theta_1}{V_1\theta}\delta\left(t - \frac{z}{V_1}\right) + \frac{\theta_2}{V_2\theta}\delta\left(t - \frac{z}{V_2}\right) . \end{aligned} \tag{D.74}$$

The first depth moment is therefore

$$\int_0^\infty z f^r(z,t)\, dz = \frac{\theta_1}{\theta}V_1 t + \frac{\theta_2}{\theta}V_2 t \tag{D.75}$$

and the mean resident velocity is given by

$$V^r = \frac{\theta_1}{\theta}V_1 + \frac{\theta_2}{\theta}V_2 = \frac{2J_w}{\theta} . \tag{D.76}$$

Solution 3.6†

If (3.61)–(3.62) are multiplied by t^N and integrated from 0 to ∞, we obtain

$$\begin{aligned} -N\theta_1 T_{N-1,1} + J_w\frac{dT_{N,1}}{dz} + \alpha T_{N,1} - \alpha T_{N,2} &= 0 , \\ -N\theta_2 T_{N-1,2} + J_w\frac{dT_{N,2}}{dz} + \alpha T_{N,2} - \alpha T_{N,1} &= 0 , \\ J_w\frac{dT_N}{dz} - N\theta_1 T_{N-1,1} - N\theta_2 T_{N-1,2} &= 0 , \end{aligned} \tag{D.77}$$

subject to the boundary conditions

$$T_{N,1}(0) = \frac{\theta_1}{\theta}\delta_{N,0} \quad , \quad T_{N,2}(0) = \frac{\theta_2}{\theta}\delta_{N,0} \quad , \quad T_N(0) = \delta_{N,0} \; , \tag{D.78}$$

where $\delta_{N,0}$ is the Kronecker δ, and

$$\begin{aligned} T_{N,i}(z) &= \int_0^\infty t^N \, C_i^f(z,t)\, dt \; ; \quad i = 1,2 \; , \\ T_N(z) &= T_{N,1}(z) + T_{N,2}(z) \; . \end{aligned} \tag{D.79}$$

Thus, for $N = 0$, $T_0 = 1$. We may write the equation for $T_{0,1}$ as

$$J_w \frac{dT_{0,1}}{dz} + 2\alpha T_{0,1} = \alpha \; , \tag{D.80}$$

since $T_{0,2} = 1 - T_{0,1}$. Equation (D.80) has the solution, using (D.78),

$$T_{0,1}(z) = \frac{1}{2}(1 - \exp(-\beta z)) + \frac{\theta_1}{\theta}\exp(-\beta z) \; , \tag{D.81}$$

where $\beta = 2\alpha/J_w$. The solution for $T_{0,2}$ is identical to (D.81) except that θ_1 has to be replaced by θ_2. Therefore, inserting the solutions for $T_{0,1}$ and $T_{0,2}$ into the last equation of (D.77), we obtain

$$J_w \frac{dT_1}{dz} = \frac{\theta}{2}(1 - \exp(-\beta z)) + \frac{\theta_1^2 + \theta_2^2}{\theta}\exp(-\beta z) = \frac{\theta}{2} + \frac{\Delta^2}{2\theta}\exp(-\beta z) \; , \tag{D.82}$$

where $\Delta = \theta_2 - \theta_1$. This has the solution

$$T_1(z) = \frac{\theta z}{2J_w} + \frac{\Delta^2}{4\alpha\theta}(1 - \exp(-\beta z)) \; , \tag{D.83}$$

Equation (D.83) is the first travel time moment of the system. For small values of α, it reduces to[2]

$$\frac{2\alpha z}{J_w} \ll 1 \quad \Rightarrow \quad T_1(z) \approx \frac{\theta z}{2J_w} + \frac{\Delta^2}{4\alpha\theta}\frac{2\alpha z}{J_w} = \frac{(\theta_1^2 + \theta_2^2)z}{\theta J_w} \; , \tag{D.84}$$

which is the result obtained in the previous problem where there was no mixing (i.e. $\alpha = 0$). For large z, it reduces to

$$\frac{2\alpha z}{J_w} \gg 1 \quad \Rightarrow \quad T_1(z) \approx \frac{\theta z}{2J_w} + \frac{\Delta^2}{4\alpha\theta} \; . \tag{D.85}$$

Therefore, in this case, the velocity V^f is twice the velocity of a mobile solute moving through a uniform medium at a water content θ, which is the same as the resident velocity (D.76) calculated in the previous problem.

[2] $\exp(-\beta z) \approx 1 - \beta z$ for $\beta z \ll 1$

The equation for the first moment of region 1 is (since $T_{1,2} = T_1 - T_{1,1}$)

$$\begin{aligned}
\frac{dT_{1,1}}{dz} + \beta T_{1,1} = \frac{\theta_1}{J_w} T_{0,1} + \frac{\alpha}{J_w} T_1 & \\
= \frac{\theta_1}{2J_w}(1 - \exp(-\beta z)) + \frac{\theta_1^2}{\theta J_w}\exp(-\beta z) & \\
+ \frac{\alpha\theta z}{2J_w^2} + \frac{\Delta^2}{4\theta J_w}(1 - \exp(-\beta z)) & \qquad \text{(D.86)} \\
= \left(\frac{\theta_1}{2J_w} + \frac{\alpha\theta z}{2J_w^2} + \frac{\Delta^2}{4\theta J_w}\right) - \left(\frac{\theta_1}{2J_w} + \frac{\Delta^2}{4\theta J_w} - \frac{\theta_1^2}{\theta J_w}\right)\exp(-\beta z) & \\
= \left(\frac{\theta_1}{2J_w} + \frac{\alpha\theta z}{2J_w^2} + \frac{\Delta^2}{4\theta J_w}\right) - \frac{\Delta}{4J_w}\exp(-\beta z) &
\end{aligned}$$

from which we obtain

$$T_{1,1} = \frac{\theta z}{4J_w} - \frac{z\Delta}{4J_w}\exp(-\beta z) - \frac{\theta_1\Delta}{4\theta\alpha}(1 - \exp(-\beta z)) \; . \qquad \text{(D.87)}$$

Similarly, the solution for $T_{1,2}$ is

$$T_{1,2} = \frac{\theta z}{4J_w} + \frac{z\Delta}{4J_w}\exp(-\beta z) + \frac{\theta_2\Delta}{4\theta\alpha}(1 - \exp(-\beta z)) \qquad \text{(D.88)}$$

Thus, the equation for T_2 becomes

$$\begin{aligned}
\frac{dT_2}{dz} &= \frac{2}{J_w}(\theta_1 T_{1,1} + \theta_2 T_{1,2}) \\
&= \frac{\theta^2 z}{2J_w^2} + \frac{z\Delta^2}{2J_w^2}\exp(-\beta z) + \frac{\Delta^2}{2\alpha J_w}(1 - \exp(-\beta z)) \; .
\end{aligned} \qquad \text{(D.89)}$$

Equation (D.89) has the solution

$$T_2 = \frac{\theta^2 z^2}{4J_w^2} + \frac{z\Delta^2}{4\alpha J_w} + \left(\frac{z\Delta^2}{4\alpha J_w} - \frac{\Delta^2}{8\alpha^2}\right)(1 - \exp(-\beta z)) \; . \qquad \text{(D.90)}$$

Thus, the travel time variance is

$$\begin{aligned}
\mathrm{Var}(t) &= T_2 - T_1^2 \\
&= \frac{z\Delta^2}{4\alpha J_w} - \frac{\Delta^2}{8\alpha^2}(1 - \exp(-\beta z)) - \frac{\Delta^4}{16\alpha^2\theta^2}(1 - \exp(-\beta z))^2 \; .
\end{aligned} \qquad \text{(D.91)}$$

For small α, this reduces to

$$\frac{2\alpha z}{J_w} \ll 1 \qquad \Rightarrow \qquad \mathrm{Var}(t) \approx \frac{\theta_1\theta_2\Delta^2 z^2}{\theta^2 J_w^2} \; , \qquad \text{(D.92)}$$

which is the result for the previous problem. For large z, the variance becomes

$$\frac{2\alpha z}{J_w} \gg 1 \qquad \Rightarrow \qquad \mathrm{Var}(t) \approx \frac{z\Delta^2}{4\alpha J_w} - \frac{\Delta^2}{8\alpha^2} - \frac{\Delta^4}{16\alpha^2\theta^2} \ . \tag{D.93}$$

Thus, the two region model converges at large times to a CDE with a constant dispersion coefficient which is proportional to the water content difference.

Solution 3.7†

The Laplace transform of the resident CDE (2.36) subject to the initial condition (3.65) and the appropriate boundary conditions is

$$s\widehat{C}_l^r - g(z) + V\frac{d\widehat{C}_l^r}{dz} - D\frac{d^2\widehat{C}_l^r}{dz^2} = 0 \ , \tag{D.94}$$

$$\widehat{C}_l^r(\infty) = 0 \ , \tag{D.95}$$

$$\left(V\widehat{C}_l^r - D\frac{d\widehat{C}_l^r}{dz}\right)\Big|_{z=0} = 0 \ . \tag{D.96}$$

If we now define the double Laplace transform as

$$\widehat{C}_l^r(r, s) := \int_0^\infty \int_0^\infty C_l^r(z, t) \exp(-rz - st)\, dz\, dt \ , \tag{D.97}$$

then the Laplace transform of (D.94) subject to (D.95)–(D.96) is

$$s\widehat{C}_l^r - \widehat{g}(r) + Vr\widehat{C}_l^r - Dr^2\widehat{C}_l^r + Dr\widehat{C}_l^r(0; s) = 0 \ , \tag{D.98}$$

where $\widehat{C}_l^r(0; s)$ is the Laplace transform with respect to time of the surface concentration $C_l^r(0, t)$, which is unknown, and $\widehat{g}(r)$ is the Laplace transform with respect to space of the initial condition (3.65). Solving for $\widehat{C}_l^r$ in (D.98), we obtain

$$\widehat{C}_l^r = \frac{Dr\widehat{C}_l^r(0, s) - \widehat{g}(r)}{Dr^2 - Vr - s} \ , \tag{D.99}$$

The denominator of (D.99) may be factored and separated into partial fractions, producing

$$\widehat{C}_l^r = \frac{1}{D(r_2 - r_1)}\left[\frac{Dr_2\widehat{C}_l^r(0, s) - \widehat{g}(r)}{r - r_2} - \frac{Dr_1\widehat{C}_l^r(0, s) - \widehat{g}(r)}{r - r_1}\right] \ , \tag{D.100}$$

where

$$r_1 := \frac{V}{2D}(1 - \xi) \quad , \quad r_2 := \frac{V}{2D}(1 + \xi) \ , \tag{D.101}$$

and ξ is given in (D.47).

These terms may now be inverted with respect to r using (C.4) from the Table of Transforms in Appendix C, substituting z for t and r for s. After inversion with respect to r, the Laplace transform $\widehat{C}_l^r(z;s)$ may be written as

$$\widehat{C}_l^r(z;s) = \frac{1}{D(r_2-r_1)}\Big[\exp(r_2 z)\Big\{Dr_2\widehat{C}_l^r(0;s) - \int_0^z g(z')\,\exp(-r_2 z')\,dz'\Big\}$$

$$- \exp(r_1 z)\Big\{Dr_1\widehat{C}_l^r(0;s) - \int_0^z g(z')\,\exp(-r_1 z')\,dz'\Big\}\Big]\ . \qquad \text{(D.102)}$$

Since $r_2 > 0$, the condition (D.95) will be met only if the coefficient of $\exp(r_2 z)$ vanishes as $z \to \infty$. Thus,

$$\widehat{C}_l^r(0;s) = \frac{1}{Dr_2}\int_0^\infty g(z')\exp(-r_2 z')\,dz'\ . \qquad \text{(D.103)}$$

Inserting (D.103) into (D.102) we obtain

$$\begin{aligned}\widehat{C}_l^r(z;s) = {}& \frac{\exp(r_2 z)}{D(r_2-r_1)}\int_z^\infty g(z')\,\exp(-r_2 z')\,dz' \\ & - \frac{\exp(r_1 z)}{D(r_2-r_1)}\Big[\frac{r_1}{r_2}\int_0^\infty g(z')\,\exp(-r_2 z')\,dz' \\ & - \int_0^z g(z')\,\exp(-r_1 z')\,dz'\Big]\ .\end{aligned} \qquad \text{(D.104)}$$

When (D.101) is substituted into (D.104), the resulting equation may be written as

$$\begin{aligned}\widehat{C}_l^r(z;s) \;=\; & \frac{1}{V\xi}\int_z^\infty g(z')\,\exp\Big(-\frac{V}{2D}(1+\xi)(z'-z)\Big)\,dz' \\ & + \frac{2}{V(1+\xi)}\int_0^\infty g(z')\,\exp\Big(-\frac{V}{2D}(z'-z) - \frac{V\xi}{2D}(z'+z)\Big)\,dz' \\ & - \frac{1}{V\xi}\int_0^\infty g(z')\,\exp\Big(-\frac{V}{2D}(z'-z) - \frac{V\xi}{2D}(z'+z)\Big)\,dz' \\ & + \frac{1}{V\xi}\int_0^z g(z')\,\exp\Big(-\frac{V}{2D}(1-\xi)(z'-z)\Big)\,dz'\ .\end{aligned} \qquad \text{(D.105)}$$

These integral forms may be inverted with (A.33) and (C.8), (C.13). The result is

$$\begin{aligned}C_l^r(z,t) \;=\; & \int_0^\infty \frac{g(z')}{2\sqrt{\pi Dt}}\,\exp\Big(-\frac{(z'-z+Vt)^2}{4Dt}\Big)\,dz' \\ & + \exp\Big(\frac{Vz}{D}\Big)\int_0^\infty \frac{g(z')}{2\sqrt{\pi Dt}}\,\exp\Big(-\frac{(z'+z+Vt)^2}{4Dt}\Big)\,dz' \\ & - \frac{V}{2D}\exp\Big(\frac{Vz}{D}\Big)\int_0^\infty g(z')\,\mathrm{erfc}\Big(\frac{z'+z+Vt}{2\sqrt{Dt}}\Big)\,dz'\ .\end{aligned} \qquad \text{(D.106)}$$

Solution 3.8

When $g(z) = \delta(z)/\theta$ is substituted into (D.106) of the previous problem, the result is

$$C_t^r(z,t) = \theta C_l^r(z,t) = f^r(z,t)$$

$$= \frac{1}{\sqrt{\pi Dt}} \exp\Big(-\frac{(z-Vt)^2}{4Dt}\Big) - \frac{V}{2D} \exp\Big(\frac{V}{2D}\Big) \operatorname{erfc}\Big(\frac{z+Vt}{2\sqrt{Dt}}\Big) . \tag{D.107}$$

This is the same as the resident pdf ((3.12) divided by J_w) of the semi-infinite CDE problem corresponding to a δ-function input of solute.

Solution 3.9

When $g(z) = H(z) - H(z-L)$ is substituted into (D.106) of problem 3.6, the result is

$$\begin{aligned} C_l^r(z,t) &= C_0\Big[\int_0^L \frac{1}{2\sqrt{\pi Dt}} \exp\Big(-\frac{(z'-z+Vt)^2}{4Dt}\Big)\, dz' \\ &\quad + \exp\Big(\frac{Vz}{D}\Big) \int_0^L \frac{1}{2\sqrt{\pi Dt}} \exp\Big(-\frac{(z'+z+Vt)^2}{4Dt}\Big)\, dz' \\ &\quad - \frac{V}{2D} \exp\Big(\frac{Vz}{D}\Big) \int_0^L \operatorname{erfc}\Big(\frac{z'+z+Vt}{2\sqrt{Dt}}\Big)\, dz'\Big] . \end{aligned} \tag{D.108}$$

The three integrals may be evaluated simply after a change of variable. Thus,

$$\begin{aligned} &\int_0^L \frac{1}{2\sqrt{\pi Dt}} \exp\Big(-\frac{(z'-z+Vt)^2}{4Dt}\Big)\, dz' \\ &\longrightarrow \frac{1}{2}\frac{2}{\sqrt{\pi}} \int_{P_1}^{P_2} \exp(-y^2)\, dy = \frac{1}{2}\big(\operatorname{erfc}(P_1) - \operatorname{erfc}(P_2)\big) , \end{aligned} \tag{D.109}$$

$$\begin{aligned} &\exp\Big(\frac{Vz}{D}\Big) \int_0^L \frac{1}{2\sqrt{\pi Dt}} \exp\Big(-\frac{(z'+z+Vt)^2}{4Dt}\Big)\, dz' \\ &\longrightarrow \frac{1}{2} \exp\Big(\frac{Vz}{D}\Big) \frac{2}{\sqrt{\pi}} \int_{P_3}^{P_4} \exp(-y^2)\, dy = \frac{1}{2} \exp\Big(\frac{Vz}{D}\Big) \big(\operatorname{erfc}(P_3) - \operatorname{erfc}(P_4)\big) , \end{aligned} \tag{D.110}$$

$$\begin{aligned} &\frac{V}{2D} \exp\Big(\frac{Vz}{D}\Big) \int_0^L \operatorname{erfc}\Big(\frac{z'+z+Vt}{2\sqrt{Dt}}\Big)\, dz' \\ &\longrightarrow \sqrt{\frac{V^2 t}{D}} \exp\Big(\frac{Vz}{D}\Big) \int_{P_3}^{P_4} \operatorname{erfc}(y)\, dy \\ &= \sqrt{\frac{V^2 t}{D}} \exp\Big(\frac{Vz}{D}\Big) \Big[P_4 \operatorname{erfc}(P_4) - P_3 \operatorname{erfc}(P_3) + \frac{1}{\sqrt{\pi}}\big(\exp(-P_3^2) - \exp(-P_4^2)\big)\Big] , \end{aligned} \tag{D.111}$$

where

$$P_1 := \frac{z-L-Vt}{2\sqrt{Dt}}\ ,\ P_2 := \frac{z-Vt}{2\sqrt{Dt}}\ ,\ P_3 := \frac{z+Vt}{2\sqrt{Dt}}\ ,\ P_4 := \frac{z+L+Vt}{2\sqrt{Dt}}\ . \tag{D.112}$$

After substitution of (D.109)–(D.112) and some simplification, (D.108) may be written as

$$\begin{aligned} C_l^r(z,t) = &\frac{C_0}{2}\Big[\operatorname{erfc}\Big(\frac{z-L-Vt}{2\sqrt{Dt}}\Big) - \operatorname{erfc}\Big(\frac{z-Vt}{2\sqrt{Dt}}\Big)\Big] \\ &-\frac{C_0}{2}\Big[\Big(1+\frac{V}{D}(z+L)+\frac{V^2t}{D}\Big)\exp\Big(\frac{Vz}{D}\Big)\operatorname{erfc}\Big(\frac{z+L+Vt}{2\sqrt{Dt}}\Big) \\ &-\Big(1+\frac{Vz}{D}+\frac{V^2t}{D}\Big)\exp\Big(\frac{Vz}{D}\Big)\operatorname{erfc}\Big(\frac{z+Vt}{2\sqrt{Dt}}\Big)\Big] \\ &+C_0\sqrt{\frac{V^2t}{\pi D}}\Big[\exp\Big(\frac{Vz}{D}-\frac{(z+L+Vt)^2}{2\sqrt{Dt}}\Big)-\exp\Big(\frac{(z-Vt)^2}{4Dt}\Big)\Big] \end{aligned} \tag{D.113}$$

This is the solution given in Van Genuchten and Alves (1982).

D.4 Problems in Chapter 4

Solution 4.1

The travel time pdf of a solute obeying (4.86) is obtained by solving it for zero initial concentration and a narrow pulse (delta function) input of solute. The Laplace transform of (4.86) is

$$(s+\mu)\widehat{C} + V\frac{d\widehat{C}}{dz} = 0 \tag{D.114}$$

subject to $\widehat{C}(0) = 1$. Thus, the solution to (D.114) is

$$\widehat{C}(z) = \exp\Big(-\frac{(s+\mu)z}{V}\Big)\ , \tag{D.115}$$

which has the inverse transform (A.4)

$$C(z,t;\mu,V) = \exp\Big(-\frac{\mu z}{V}\Big)\,\delta\Big(t-\frac{z}{V}\Big) = \exp(-\mu t)\,\delta\Big(t-\frac{z}{V}\Big)\ . \tag{D.116}$$

This is the travel time pdf of the stream tube. It is not normalized to unit area because not all of the solute reaches the outflow end. For constant V, the travel time pdf of the transport volume is obtained by inserting (4.87) into (4.1), and integrating over the range of μ values in the transport volume. Thus, using (D.39) we obtain

$$\begin{aligned} \overline{C}(z,t) &= E\big(C(z,t;\mu,V)\big) = \delta\Big(t-\frac{z}{V}\Big)\int_0^\infty \exp(-\mu t)\,f_\mu(\mu)\,d\mu \\ &= \delta\Big(t-\frac{z}{V}\Big)\frac{\beta^{N+1}}{N!}\int_0^\infty \exp(-\mu(t+\beta))\,\mu^N\,d\mu = \delta\Big(t-\frac{z}{V}\Big)\Big(1+\frac{t}{\beta}\Big)^{-(N+1)}\ . \end{aligned} \tag{D.117}$$

This is no longer a first-order decay process.

If μ is constant and V is variable, then the pdf of the transport volume is

$$\begin{aligned}\overline{C}(z,t) &= E\big(C(z,t;\mu,V)\big) = \exp(-\mu t)\int_0^\infty \delta\Big(t-\frac{z}{V}\Big)\, f_V(V)\, dV \\ &= \exp(-\mu t)\int_0^\infty \delta(t-y) f_V\Big(\frac{z}{y}\Big)\frac{z}{y^2}\, dy = \exp(-\mu t)\,\frac{z}{t^2}\, f_V\Big(\frac{z}{t}\Big)\,.\end{aligned} \tag{D.118}$$

This decay process is still first-order, regardless of the distribution of V.

Solution 4.2

The mean travel time pdf (4.34) may be integrated over t_m, producing

$$f^f(\ell,t) = \int_0^\infty\int_0^\infty \delta(t-Rt_m)\, f(R,t_m)\, dR dt_m = \int_0^\infty f\Big(R,\frac{t}{R}\Big)\frac{dR}{R}\,, \tag{D.119}$$

where $f(R,t_m)$ is the joint pdf of R and t_m. In general, if R and t_m are lognormally distributed, then the bivariate lognormal distribution is written as

$$f(R,t_m) = \frac{1}{2\pi\,\sigma_R\,\sigma_m\, R\, t_m\sqrt{1-\rho^2}}\exp\Big(-\frac{(y_R^2+y_m^2-2\rho\, y_R\, y_m)}{2(1-\rho^2)}\Big)\,, \tag{D.120}$$

where

$$y_R = \frac{\ln(R)-\mu_R}{\sigma_R}\quad,\quad y_m = \frac{\ln(t_m)-\mu_m}{\sigma_m}\,. \tag{D.121}$$

After insertion of (D.120) into (D.119) and a change of variable to $u=\ln(R)$, we obtain

$$f^f(\ell,t) = \frac{1}{2\pi\,\sigma_R\sigma_m\, t\sqrt{1-\rho^2}}\int_{-\infty}^\infty \exp(-Au^2-Bu-C)\, du\,, \tag{D.122}$$

where

$$\begin{aligned} A &= \frac{1}{2(1-\rho^2)}\Big(\frac{1}{\sigma_R^2}+\frac{1}{\sigma_m^2}+\frac{2\rho}{\sigma_R\,\sigma_m}\Big)\,, \\ B &= -\frac{1}{1-\rho^2}\Big(\frac{\mu_R}{\sigma_R^2}+\frac{\mu_\beta}{\sigma_m^2}+\frac{\rho(\mu_R+\mu_\beta)}{\sigma_R\,\sigma_m}\Big)\,, \\ C &= \frac{1}{2(1-\rho^2)}\Big(\frac{\mu_R^2}{\sigma_R^2}+\frac{\mu_\beta^2}{\sigma_m^2}+\frac{2\rho\,\mu_R\,\mu_\beta}{\sigma_R\,\sigma_m}\Big)\,,\end{aligned} \tag{D.123}$$

where $\mu_\beta = \ln(t)-\mu_m$. Since

$$\int_{-\infty}^{+\infty}\exp\big(-(Ax^2+Bx+C)\big)\, dx = \sqrt{\frac{\pi}{A}}\exp\Big(\frac{B^2-4AC}{4A}\Big) \tag{D.124}$$

(Abramowitz and Stegun, 1970), after some simplification, (D.122) reduces to

$$f^f(\ell,t) = \frac{1}{\sqrt{2\pi}\sigma_t t}\exp\Big(-\frac{(\ln(t)-\mu_t)^2}{2\sigma_t^2}\Big) \tag{D.125}$$

where

$$\mu_t = \mu_R + \mu_m \quad , \quad \sigma_t^2 = \sigma_R^2 + \sigma_m^2 + 2\rho\,\sigma_R\,\sigma_m \ . \tag{D.126}$$

Thus, t is lognormally distributed.

Solution 4.3

From the previous problem, it is clear that if I and i are lognormally distributed random variables, then $t = I/i$ is lognormally distributed with parameters that obey

$$\mu_t = \mu_I - \mu_i \quad , \quad \sigma_t^2 = \sigma_I^2 + \sigma_i^2 - 2\rho\,\sigma_I\,\sigma_i \ . \tag{D.127}$$

where ρ is the correlation coefficient of the log-variates of I and i. If we reason that the portions of the field which have the highest infiltration rate are also the portions of the field requiring the least applied water to leach solutes to a given depth, then I and i will be inversely correlated and $\rho = -1$.

With this assumption, $\mu_I = 2.00$ and $\sigma_I = 0.50$. These values are close to those found for soils of similar texture for field studies of solute leaching under unsaturated leaching by sprinkler or trickle irrigation (Jury, 1985).

Solution 4.4

Since $Y = aX^b$ then $\ln(Y) = \ln(a) + b\ln(X)$. If we use the following transformation for pdfs (Grimmet and Welch, 1986)

$$f_y(y) = f_x(x(y))\frac{dx}{dy} \tag{D.128}$$

which is valid for any continuous function $x(y)$, then

$$\begin{aligned} f_Y(Y) &= \frac{1}{\sqrt{2\pi}\sigma_X}\exp\left[-\frac{\left((\ln(Y)-\ln(a))/b-\mu_X\right)^2}{2\sigma_X^2}\right]\frac{1}{bY} \\ &= \frac{1}{\sqrt{2\pi}\{b\sigma_X\}Y}\exp\left(-\frac{\left(\ln(Y)-\{b\mu_X+\ln(a)\}\right)^2}{2\{b\sigma_X\}^2}\right) . \end{aligned} \tag{D.129}$$

Thus, Y is lognormally distributed with $\mu_Y = b\mu_X + \ln(a)$ and $\sigma_Y = b\sigma_X$.

Solution 4.5

Application of (3.37) to the flux pdf (4.4) results in, after changing the integration variable to $y = z/t$,

$$f_t^r(z) = -\frac{\partial}{\partial z}\int_0^t \frac{z}{t^2} f_v\left(\frac{z}{t}\right) dt = \frac{\partial}{\partial z}\int_\infty^{z/t} f_V(y)\,dy = \frac{1}{t} f_V\left(\frac{z}{t}\right) , \tag{D.130}$$

where we have used Leibnitz's rule (D.55) for the differentiation of the integral.

Solution 4.6

Here the stream tube problem must be solved for the flux concentration input condition

$$J_s(0,t) = iC^f(0,t) = C_0 i\big(H(t) - H(t-\Delta t)\big) \ . \tag{D.131}$$

The mass added to a given stream tube is given by

$$M = \int_0^\infty J_s(0,t)\,dt = C_0 i \Delta t \ . \tag{D.132}$$

Therefore, the mass is not constant everywhere since i is variable. Rather,

$$\mathrm{E}(M) = C_0 \Delta t\,\mathrm{E}(i) \quad , \quad \mathrm{Var}(M) = C_0 \Delta t\,\mathrm{Var}(i) \ . \tag{D.133}$$

Thus, those stream tubes which have larger i will contribute more solute molecules to the determination of the field scale pdf than the tubes with lower i.

Solution 4.7

Here we use (4.49)

$$\mathrm{E}_z(t^N) = \int_0^\infty \int_0^\infty \mathrm{E}_z(t^N; V, D)\, f_V(V)\, f_D(D)\, dV dD \ . \tag{D.134}$$

The mean and variance of the local CDE process are given by (2.57)–(2.58). Therefore, inserting these into (D.134), we obtain

$$\begin{aligned} \mathrm{E}_z(t) &= z\,\mathrm{E}_V\Big(\frac{1}{V}\Big) \ , \\ \mathrm{Var}_z(t) &= 2z\,\mathrm{E}_D(D)\,\mathrm{E}_V\Big(\frac{1}{V^3}\Big) + z^2\,\mathrm{Var}_V\Big(\frac{1}{V}\Big) \ . \end{aligned} \tag{D.135}$$

Using (2.77), we may express the result in terms of the lognormal parameters of $f_V(V)$ and $f_D(D)$,

$$\begin{aligned} \mathrm{E}_z(t) &= z\exp\Big(-\mu_V + \frac{\sigma_V^2}{2}\Big) \ , \\ \mathrm{Var}_z(t) &= 2z\exp\Big(\mu_D - 3\mu_V + \frac{\sigma_D^2}{2} + \frac{9\sigma_V^2}{2}\Big) + z^2\exp(-2\mu_V + \sigma_V^2)\,(\exp(\sigma_V^2) - 1) \ . \end{aligned} \tag{D.136}$$

If we define the macroscopic dispersion coefficient and mean velocity in terms of the moments of the mean flux pdf using (2.57)–(2.58), then

$$\begin{aligned} \frac{1}{V^f} &= \mathrm{E}_V\Big(\frac{1}{V}\Big) = \exp\Big(-\mu_V + \frac{\sigma_V^2}{2}\Big) \ , \\ D^f &= \frac{(V^f)^3}{2z}\mathrm{Var}_z(t) \\ &= \frac{\mathrm{E}_D(D)\,\mathrm{E}_V(V^{-3})}{\mathrm{E}_V^3(V^{-1})} + \frac{z}{2}\frac{\mathrm{Var}_V(V^{-1})}{\mathrm{E}_V^3(V^{-1})} \\ &= \exp\Big(\mu_D + \frac{\sigma_D^2}{2} + 3\sigma_V^2\Big) + \frac{z}{2}\exp\Big(\mu_V - \frac{\sigma_V^2}{2}\Big)\,(\exp(\sigma_V^2) - 1) \ . \end{aligned} \tag{D.137}$$

Thus, the macrodispersion coeffiecient has a linearly increasing term which is independent of the local dispersion process, and a constant term which depends on the local D and V.

Solution 4.8†

The problem to be solved may be stated as

$$\begin{aligned} \theta\frac{\partial C_l^r}{\partial t} + \rho_b\frac{\partial C_a^r}{\partial t} + \theta V\frac{\partial C_l^r}{\partial z} - \theta D\frac{\partial^2 C_l^r}{\partial z^2} &= 0 , \\ \rho_b\frac{\partial C_a^r}{\partial t} - \alpha(K_d C_l^r - C_a^r) &= 0 , \end{aligned} \tag{D.138}$$

subject to

$$\begin{aligned} VC_l^r - D\frac{\partial C_l^r}{\partial z} &= 0 \qquad \text{at } z = \pm\infty , \\ C_l^r(z,0) &= \delta(z) . \end{aligned} \tag{D.139}$$

If we multiply (D.138)–(D.139) by z^N and integrate from $-\infty$ to ∞, we obtain

$$\theta\frac{dZ_{NL}}{dt} + \rho_b\frac{dZ_{NA}}{dt} - N\theta V Z_{N-1,L} - N(N-1)\theta D Z_{N-2,L} = 0 , \tag{D.140}$$

$$\rho_b\frac{dZ_{NA}}{dt} = \alpha(K_d Z_{NL} - Z_{NA}) , \tag{D.141}$$

subject to the initial conditions

$$Z_{NL}(0) = \frac{1}{R\theta}\delta_{N,0} \tag{D.142}$$

$$Z_{NA}(0) = \frac{K_d}{R\theta}\delta_{N,0} \tag{D.143}$$

where

$$Z_{NL}(t) = \int_{-\infty}^{\infty} z^N C_l^r(z,t)\, dz \tag{D.144}$$

is the Nth depth moment. Equations (D.142)–(D.143) represent the assumed equilibrium condition at $t = 0$.

For $N = 0$ the solution to (D.140)–(D.141) is

$$Z_{0L}(t) = \frac{1}{R\theta} , \tag{D.145}$$

$$Z_{0A}(t) = \frac{K_d}{R\theta} . \tag{D.146}$$

Thus, since the phases have an equilibrium mass balance at time 0, they maintain it for all time.[3]

[3] For any initial condition other than (D.142)–(D.143), the $N = 0$ moments of the phases would be time-dependent.

For $N = 1$, (D.140) may be integrated with respect to time, producing

$$Z_{1T}(t) := \theta Z_{1L}(t) + \rho_b Z_{1A}(t) = \frac{Vt}{R\theta} \ . \tag{D.147}$$

This result may be inserted into (D.141) and solved for $Z_{1L}(t)$, producing

$$Z_{1L}(t) = \frac{Vt}{R^2\theta} + \frac{V\rho_b}{\alpha R^2\theta}\Big(1 - \frac{1}{R}\Big)\Big(1 - \exp\Big(-\frac{\alpha Rt}{\rho_b}\Big)\Big) \ . \tag{D.148}$$

Thus, the total resident concentration has a constant velocity. However, the velocity of the dissolved solute decreases from an initial value that is the same as the water,[4] to a final value that is equal to V/R. Therefore, if the progress of a chemical pulse was monitored by occasional soil coring, the velocity would be constant, but if it was monitored by resident solution sampling (by an infinitely fast extraction device), the velocity would apparantly decrease over time.

Solution 4.9

To evaluate the Nth moment, we use (2.29) on the Laplace transform of the flux pdf given by (4.30)

$$\widehat{f}^f(z;s) \;=\; \exp\Big(\frac{Vz}{2D}(1-\zeta)\Big) \ , \tag{D.149}$$

$$\zeta \;=\; \sqrt{1 + \frac{4g(s)D}{V}} \ , \tag{D.150}$$

$$g(s) \;=\; s + \frac{s\beta(R-1)}{s+\beta} \ , \tag{D.151}$$

where $\beta := \alpha/\rho_b$. By the chain rule of differentiation, we may write

$$\frac{d\widehat{f}^f}{ds} = \frac{dg}{ds}\frac{d\zeta}{dg}\frac{d\widehat{f}^f}{d\zeta} \ , \tag{D.152}$$

where

$$\frac{d\widehat{f}^f}{d\zeta} \;=\; -\frac{Vz}{2D}\widehat{f}^f \;\xrightarrow{s\to 0}\; -\frac{Vz}{2D} \ , \tag{D.153}$$

$$\frac{d\zeta}{dg} \;=\; \frac{2D}{V^2\zeta} \;\xrightarrow{s\to 0}\; \frac{2D}{V^2} \ , \tag{D.154}$$

$$\frac{dg}{ds} \;=\; 1 + \frac{\beta(R-1)}{s+\beta} - \frac{\beta s(R-1)}{(s+\beta)^2} \;\xrightarrow{s\to 0}\; R \ . \tag{D.155}$$

Thus,

$$-\frac{d\widehat{f}^f}{ds} = \frac{z}{V\zeta}\frac{dg}{ds}\widehat{f}^f \;\xrightarrow{s=0}\; \mathrm{E}_z(t) = \frac{Rz}{V} \ . \tag{D.156}$$

[4]This may be shown by evaluating (D.148) for small time.

Similarly,

$$\frac{d^2 \widehat{f}^f}{ds^2} = \left[\frac{z}{V\zeta^2} \frac{d\zeta}{dg} \left(\frac{dg}{ds}\right)^2 + \left(\frac{z}{V\zeta}\right)^2 \left(\frac{dg}{ds}\right)^2 - \frac{z}{V\zeta} \frac{d^2 g}{ds^2} \right] \widehat{f}^f , \tag{D.157}$$

where

$$\frac{d^2 g}{ds^2} = -\frac{2\beta(R-1)}{(s+\beta)^2} + \frac{2\beta s(R-1)}{(s+\beta)^3} \quad \overset{s\to 0}{\longrightarrow} \quad -\frac{2(R-1)}{\beta} . \tag{D.158}$$

Thus, after setting $s = 0$ in (D.157), we obtain

$$\mathrm{E}_z(t^2) = \frac{2zDR^2}{V^3} + \frac{z^2 R^2}{V^2} + \frac{2z\rho_b(R-1)}{\alpha V} , \tag{D.159}$$

and, with (D.156)

$$\mathrm{Var}_z(t) = \frac{2zDR^2}{V^3} + \frac{2z\rho_b(R-1)}{\alpha V} . \tag{D.160}$$

D.5 Problems in Chapter 5

Solution 5.1

The flux pdf is given by (5.8), and the relation between flux and resident pdfs by (3.37). Therefore,

$$f_t^r(z) = -\frac{\partial}{\partial z} \int_0^t f^f(z,t')\, dt' = -\frac{dy}{dz}\frac{\partial}{\partial y} \int_0^t f^f(z,t')\, dt' = -\theta(z)\frac{\partial}{\partial y} \int_0^t f^f(z,t')\, dt' . \tag{D.161}$$

The flux pdf (5.8) is identical to the CDE flux pdf (2.51), except that

$$z \to y(z) \quad , \quad V \to J_w \quad , \quad D \to E . \tag{D.162}$$

Therefore, the resident pdf above is equal to $\theta(z)$ times the resident pdf of the CDE with the above variable substitutions.

Solution 5.2

The attribute of the stochastic-convective pdf we wish to maintain in heterogeneous soil is the condition that the probability of reaching a depth z when an amount of water less than or equal to I has been added to the surface should be equal to the probability of reaching a depth ℓ when an amount of water less than or equal to $Iy(\ell)/y(z)$ has been added to the surface, or

$$P(z,I) = P(\ell, I\, y(\ell)/y(z)) . \tag{D.163}$$

This adjusts the earlier postulate (2.64) for the variable amount of mean storage in different parts of the transport volume. Since the travel time pdf in terms of I is the derivative of the cdf with respect to I, we obtain

$$f^f(z,I) = \frac{dP(z,I)}{dI} = \frac{dP(\ell, I\, y(\ell)/y(z))}{dI} = \frac{y(\ell)}{y(z)}\, f^f\left(\ell, I\frac{y(\ell)}{y(z)}\right) . \tag{D.164}$$

Solution 5.3

The mean is clearly given by (5.20), because the ensemble average is a linear operator. The second moment of Z can be written as

$$\mathrm{E}(Z^2) = \mathrm{E}\left[\sum_{i=1}^{N} Z_i \sum_{j=1}^{N} Z_j\right] = \sum_{i=1}^{N}\sum_{j=1}^{N} \mathrm{E}(Z_i Z_j) . \tag{D.165}$$

We now write

$$\mathrm{E}(Z_i Z_j) = \mu_i \mu_j + \rho_{ij}\, \sigma_i\, \sigma_j , \tag{D.166}$$

which follows from (5.18). Then, since

$$\mathrm{Var}(Z) = \mathrm{E}(Z^2) - \mathrm{E}^2(Z) , \tag{D.167}$$

it follows that (5.21) is valid.

Solution 5.4†

From (5.12), (5.25), and (5.31), we may write the flux pdf at $z = 2\ell$ as

$$f^f(2\ell, t) = \int_0^\infty \int_0^\infty \delta(t - t_1 - t_2)\, f_1^f(\ell, t_1)\, \delta(t_2 - \alpha t_1^\beta)\, dt_1 dt_2 , \tag{D.168}$$

where $\beta := \sigma_2/\sigma_1$ and $\alpha := \exp(\mu_2 - \mu_1\sigma_2/\sigma_1)$. Thus, integrating over t_2, we obtain

$$f^f(2\ell, t) = \int_0^\infty \delta(t - t_1 - \alpha t_1^\beta)\, f_1^f(\ell, t_1)\, dt_1 . \tag{D.169}$$

If we define

$$g(t_1) := t_1 + \alpha t_1^\beta \tag{D.170}$$

then the integral may be evaluated by (2.10)

$$f^f(2\ell, t) = f_1^f(\ell, g^{-1}(t)) \Big/ \frac{dg}{dt_1}(g^{-1}(t)) , \tag{D.171}$$

where $g^{-1}(t)$ is the x root of the equation

$$t - x - \alpha x^\beta = 0 , \tag{D.172}$$

which may be found rapidly for a given t by the Newton-Raphson procedure (Press et al., 1986) or some other method. Note that

$$\frac{dg}{dt_1} = 1 + \beta\alpha t_1^{\beta-1} . \tag{D.173}$$

Thus, letting $t^* = g^{-1}(t)$, Equation (D.171) may be written

$$f^f(2\ell, t) = \frac{f_1^f(\ell, t^*)}{1 + \beta\alpha t^{*\beta-1}} , \tag{D.174}$$

where

$$t - t^* - \alpha t^{*\beta} = 0 , \tag{D.175}$$

and $f_1^f(\ell, t^*)$ is the lognormal travel time pdf of layer 1 evaluated at $t = t^*$.

Solution 5.5†

The resident pdf is calculated with (3.37) applied to (D.168). Thus, we obtain, from (3.44),

$$f^r(z,t) = \frac{t}{z} f_1^f(z,t) \; ; \qquad z < \ell \; . \tag{D.176}$$

Below $z = \ell$, the thickness of the lower layer is $z - \ell$, and the relationship in (5.30) becomes

$$\frac{\ln(t_1) - \mu_1}{\sigma_1} = \frac{\ln\left(t_2\ell/(z-\ell)\right) - \mu_2}{\sigma_2} \; , \tag{D.177}$$

where μ and σ in each zone are defined with respect to a reference thickness ℓ. Thus,

$$t_2 = \frac{z-\ell}{\ell} \exp\left(\mu_2 - \mu_1 \frac{\sigma_2}{\sigma_1}\right) t_1^{\sigma_2/\sigma_1} = \alpha(z)\, t_1^\beta \; . \tag{D.178}$$

Therefore, the resident pdf for $z > \ell$ is

$$f^r(z,t) = -\frac{\partial}{\partial z} \int_0^t \int_0^\infty \int_0^\infty \delta(t' - t_1 - t_2)\, f_1^f(\ell, t_1)\, \delta(t_2 - \alpha(z) t_1^\beta)\, dt_1 dt_2 dt' \; ; \quad z > \ell \; . \tag{D.179}$$

Integrating over t' and t_2 we obtain, by (2.14)

$$\begin{aligned} f^r(z,t) &= -\frac{\partial}{\partial z} \int_0^\infty H(t - t_1 - \alpha(z) t_1^\beta)\, f_1^f(\ell, t_1)\, dt_1 \\ &= \int_0^\infty \delta(t - t_1 - \alpha(z) t_1^\beta)\, \frac{d\alpha(z)}{dz}\, t_1^\beta\, f_1^f(\ell, t_1)\, dt_1 \; . \end{aligned} \tag{D.180}$$

Thus, following the same procedure as in the last problem, we have

$$f^r(z,t) = \frac{d\alpha(z)}{dz}\, t^{*\beta}\, f_1^f(\ell, t^*) \Big/ \left(1 + \beta\alpha(z) t^{*\beta - 1}\right) \; , \tag{D.181}$$

where t^* is the root of

$$t - t^* - \alpha(z) t^{*\beta} = 0 \; . \tag{D.182}$$

Solution 5.6

The resident fluid concentration is equal to, by (3.2)

$$C_l^r(z,t) = -\int_0^t V \frac{\partial C^f(z,t)}{\partial z}\, dt \; . \tag{D.183}$$

Thus, the condition of continuous resident concentration is equivalent to

$$0 = C_{1l}^r(\ell,t) - C_{2l}^r(\ell,t) = -\int_0^t \left(V \frac{\partial C_1^f}{\partial z} - V \frac{\partial C_2^f}{\partial z}\right) dt \tag{D.184}$$

for all t. Therefore, the argument of the integral must vanish.

Solution 5.7

By (2.68), the variance in region 1 is equal to

$$\mathrm{Var}_z(t) = \left(\frac{z}{\ell}\right)^2 \mathrm{Var}_\ell(t) \; ; \qquad 0 < z < \ell \; , \tag{D.185}$$

since region 1 is stochastic-convective. Since the two layers are identical, but uncorrelated, for depths $z > \ell$, the mean and variance are

$$\mathrm{E}_z(t) = \frac{z}{\ell}\,\mathrm{E}_\ell(t) \; ; \qquad z > 0 \; , \tag{D.186}$$

$$\mathrm{Var}_z(t) = \mathrm{Var}_\ell(t) + \mathrm{Var}_{z-\ell}(t) \; ; \qquad 0 < z < \ell \; . \tag{D.187}$$

However, the second layer of thickness $z - \ell$ is also stochastic-convective, and therefore with respect to the reference thickness ℓ it obeys

$$\mathrm{Var}_{z-\ell}(t) = \left(\frac{z-\ell}{\ell}\right)^2 \mathrm{Var}_\ell(t) \; , \tag{D.188}$$

so that (D.187) becomes

$$\mathrm{Var}_z(t) = \left(1 + \left(\frac{z-\ell}{\ell}\right)^2\right) \mathrm{Var}_\ell(t) \; ; \qquad z > \ell \; . \tag{D.189}$$

The generalized definition of the dispersivity λ is (2.57)– (2.58),

$$\lambda := \frac{D}{V} = \frac{z}{2}\frac{\mathrm{Var}_z(t)}{\mathrm{E}_z^2(t)} \; . \tag{D.190}$$

Thus, after inserting the above expressions into (D.190), we may write the apparent dispersivity as

$$\frac{\lambda_z}{\lambda_\ell} = \begin{cases} \dfrac{z}{\ell} & , 0 < z < \ell \; ; \\ \dfrac{\ell}{z}\left(1 + \left(\dfrac{z-\ell}{\ell}\right)^2\right) & , z > \ell \; . \end{cases} \tag{D.191}$$

The apparent dispersivity thus decreases from λ_ℓ to $0.83\lambda_\ell$ between ℓ and $\sqrt{2}\ell$ and then begins increasing again, reaching a value of λ_ℓ at $z = 2\ell$.

Solution 5.8

The variance in region 1 is unchanged from the previous problem. Furthermore, the variance for all z obeys

$$\mathrm{Var}_z(t) = \left(\frac{z}{\ell}\right)^2 \mathrm{Var}_\ell(t) \; , \tag{D.192}$$

since the variance of each layer is the same per unit length, and the layers are perfectly correlated. However, for this problem, the mean travel time obeys

$$\mathrm{E}_z(t) = \begin{cases} \dfrac{z}{\ell}\,\mathrm{E}_\ell(t) & , 0 < z < \ell \; ; \\ \left(1 + 2\dfrac{z-\ell}{\ell}\right)\mathrm{E}_\ell(t) & , \ell < z < 2\ell \; ; \\ \dfrac{z+\ell}{\ell}\,\mathrm{E}_\ell(t) & , z > 2\ell \; . \end{cases} \tag{D.193}$$

In this case, the apparant dispersivity has the following scale dependence on z

$$\frac{\lambda_z}{\lambda_\ell} = \begin{cases} \dfrac{z}{\ell} & , 0 < z < \ell \, ; \\ \dfrac{(z/\ell)^3}{\left(2z/\ell - 1\right)^2} & , \ell < z < 2\ell \, ; \\ \dfrac{(z/\ell)^3}{(z+\ell)^2} & , z > 2\ell \, . \end{cases} \tag{D.194}$$

This expression decreases in the zone $\ell < z < 2\ell$ to a minimum of 0.84 and then increases again.

The significance of the results from these two problems is that they produce about the same structure for the dispersivity change using two very different assumptions about heterogeneity and interface correlation. In the first case, the termination of the correlation structure at the interface would probably be the result of enhanced lateral mixing, due to subsurface runoff. In the second case there was no mixing at the interface, merely a change in the mean layer properties.

Many more variations are possible by allowing each region to have different means and variances. Thus, the variance structure of the travel time pdf is not sufficient to explore correlations at interfaces in heterogeneous soils.

Solution 5.9

To show discontinuity, we will evaluate the resident concentration on each side of the interface. On the upper side, we obtain, from (D.51)

$$\widehat{C}_l^r(z;s) \longrightarrow \frac{2}{1+\xi_1} \exp\left(-\frac{V_1\ell(1-\xi_1)}{2D_1}\right) \; ; \qquad z \uparrow \ell \, , \tag{D.195}$$

where $\xi_1 := \sqrt{1 + 4sD_1/V_1^2}$. On the lower side, we obtain, from (2.50), (3.7), and (5.24)

$$\widehat{C}_l^r(z;s) \longrightarrow \frac{2}{1+\xi_2} \exp\left(-\frac{V_1\ell(1-\xi_1)}{2D_1}\right) \; ; \qquad z \downarrow \ell \, , \tag{D.196}$$

where $\xi_2 := \sqrt{1 + 4sD_2/V_2^2}$. Therefore, there will be a discontinuity in the resident concentration at $z = \ell$ unless $\xi_1 = \xi_2$, which is only valid in homogeneous soil.

Solution 5.10†

Since the partial differential equation for the flux concentration in each region is a CDE, the Laplace transform of the general solution in each region is given by (2.49)

$$\widehat{C}_{1L}^f(z;s) = A_1 \exp\left(\frac{V_1 z}{2D_1}(1-\xi_1)\right) + B_1 \exp\left(\frac{V_1 z}{2D_1}(1+\xi_1)\right) \; ; \qquad 0 \le z < L_1 \, , \tag{D.197}$$

$$\widehat{C}_{2L}^f(z;s) = A_2 \exp\left(\frac{V_2 z}{2D_2}(1-\xi_2)\right) + B_2 \exp\left(\frac{V_2 z}{2D_2}(1+\xi_2)\right) \; ; \qquad z > L_1 \, , \tag{D.198}$$

where $\xi_i := \sqrt{1+4sD_i/V_i^2}$, $i = 1,2$. Since we are solving for the flux pdf, the upper boundary has a $\delta(t)$ inlet flux condition. Therefore,

$$\widehat{C}^f_{1L}(0;s) = A_1 + B_1 = 1 \ . \tag{D.199}$$

The solution must be finite at $z = \infty$, so that

$$\widehat{C}^f_{2L}(\infty;s) = B_2 = 0 \ . \tag{D.200}$$

Continuity of flux and resident concentrations (5.39)–(5.40) require that

$$\begin{aligned} A_1\omega_1 + (1-A_1)\omega_2 &= A_2\omega_3 \ , \\ A_1(1-\xi_1)\omega_1 + (1-A_1)(1+\xi_1)\omega_2 &= A_2\chi(1-\xi_2)\omega_3 \ , \end{aligned} \tag{D.201}$$

where

$$\omega_1 := \exp\Big(\frac{V_1L_1}{2D_1}(1-\xi_1)\Big) \ , \ \omega_2 := \exp\Big(\frac{V_1L_1}{2D_1}(1+\xi_1)\Big) \ , \ \omega_3 := \exp\Big(\frac{V_2L_1}{2D_2}(1-\xi_2)\Big) \ , \tag{D.202}$$

$$\chi := \frac{V_2^2D_1}{V_1^2D_2} \ . \tag{D.203}$$

Thus, after elimination of A_2 in (D.201), we obtain

$$\begin{aligned} A_1 &= \frac{((1+\xi_1) - \chi(1-\xi_2))\omega_2}{\chi(1-\xi_2)(\omega_1-\omega_2) + (1+\xi_1)\omega_2 - (1-\xi_1)\omega_1} \ , \\ B_1 &= 1 - A_1 \ , \\ A_2 &= \frac{A_1(\omega_1-\omega_2)+\omega_2}{\omega_3} \ . \end{aligned} \tag{D.204}$$

Solution 5.11

In the surface layer, the flux pdf is given by (4.67)

$$f^f(z,t) = \exp(-\mu t) f^f_m(z,t) \ , \tag{D.205}$$

where $f^f_m(z,t)$ is the stochastic-convective pdf of the mobile, nondecaying chemical.

At a depth $z > \ell$, the travel time t is given by

$$f^f(z,t) = \int_0^\infty \delta\Big(t - \frac{zt'}{\ell}\Big) \ \exp(-\mu t') \ f^f_m(z,t')\, dt' \ , \tag{D.206}$$

where the condition $t = zt'/\ell$ imposed by the delta function represents the stochastic-convective hypothesis, and $\exp(-\mu t')\ f^f_m(z,t')$ is the travel time pdf of t' in the region $0 < z < \ell$. Thus, for $z > \ell$

$$f^f(z,t) = \frac{\ell}{z} \ \exp\Big(-\mu t\frac{\ell}{z}\Big) \ f^f_m\Big(z, t\frac{\ell}{z}\Big) \ . \tag{D.207}$$

D.6 Problems in Chapter 6

Solution 6.1

We first insert the scaling relations (6.27)–(6.28) into the Richards equation (6.42). Assuming that the α_i are constant at a local site, we may write the new equation as

$$\frac{\partial \theta}{\partial t} = \alpha_i \frac{\partial}{\partial z}\left(K^*(\theta)\frac{\partial \psi^*}{\partial z}\right) - \alpha_i^2 \frac{\partial}{\partial z} K^*(\theta) \ . \tag{D.208}$$

Next, we define scaled time and space variables $T := t/\tau$, $X := z/\zeta$, and then substitute into (D.208)

$$\frac{\partial \theta}{\partial T} = \frac{\tau \alpha_i}{\zeta^2}\frac{\partial}{\partial X}\left(K^*(\theta)\frac{\partial \psi^*}{\partial X}\right) - \frac{\tau \alpha_i^2}{\zeta}\frac{\partial}{\partial X} K^*(\theta) \ . \tag{D.209}$$

Finally, we require that

$$\frac{\tau \alpha_i}{\zeta^2} = 1 \quad , \quad \frac{\tau \alpha_i^2}{\zeta} = 1 \ , \tag{D.210}$$

which has the solution

$$\begin{aligned} \tau &= \alpha_i^{-3} \quad &\Rightarrow \quad T &= \alpha_i^3 t \ , \\ \zeta &= \alpha_i^{-1} \quad &\Rightarrow \quad X &= \alpha_i z \ . \end{aligned} \tag{D.211}$$

Therefore, one need only solve the equation

$$\frac{\partial \theta}{\partial T} = \frac{\partial}{\partial X}\left(K^*(\theta)\frac{\partial \psi^*}{\partial X}\right) - \frac{\partial}{\partial X} K^*(\theta) \tag{D.212}$$

once for a given set of boundary and initial conditions (assuming that they can also be scaled), from which the solution $\theta(X,T)$ can be transformed to a local solution $\theta_i(z,t)$ using (D.211). This derivation and its application was published by Warrick and Amoozegar-Fard (1982).

Solution 6.2

The relation between α and V is given in (6.35). By inserting this into (6.37) we obtain

$$f_V(V) = \begin{cases} \dfrac{1}{2\beta V} f_Y\left(\dfrac{\ln(V) - \gamma}{2\beta}\right) & , \ln(V) > 2\beta Y_P + \gamma \ , \\ \dfrac{1}{2V} f_Y\left(\dfrac{\ln(V) - \eta}{2}\right) & , \ln(V) \le 2Y_P + \eta \ , \end{cases} \tag{D.213}$$

where γ, β, and Y_P are defined in (6.36). From these definitions it is easy to show that

$$2\beta Y_P + \gamma = 2Y_P + \eta = \ln\left(\frac{i_0}{\theta_s}\right) \ . \tag{D.214}$$

Therefore

$$\int_0^\infty f_V(V)\,dV = \int_{-\infty}^\infty f_V(\ln(V))\,d\ln(V)$$

$$= \int_{-\infty}^{\ln(i_0/\theta_s)} f_Y((x-\eta)/2)\,\frac{dx}{2} + \int_{\ln(i_0/\theta_s)}^\infty f_Y((x-\gamma)/2\beta)\,\frac{dx}{2\beta} \tag{D.215}$$

$$= \int_{-\infty}^{(\ln(i_0/\theta_s)-\eta)/2} f_Y(\xi)\,d\xi + \int_{(\ln(i_0/\theta_s)-\gamma)/2\beta}^\infty f_Y(\xi)\,d\xi = 1\ ,$$

since

$$\frac{\ln(i_0/\theta_s)-\eta}{2} = \frac{\ln(i_0/\theta_s)-\gamma}{2\beta} = \frac{\ln(i_0/K_s^*)}{2}\ . \tag{D.216}$$

Therefore, $f_Y(Y)$ is normalized.

Solution 6.3

In groundwater, if the porosity ϕ is constant, we may write the CDE equation as

$$\frac{\partial C^f}{\partial t} + \frac{J_w}{\phi}\frac{\partial C^f}{\partial z} - D\frac{\partial^2 C^f}{\partial z^2} = 0\ . \tag{D.217}$$

If we now let $I = J_w t$, (D.217) becomes

$$\frac{\partial C^f}{\partial I} + \frac{1}{\phi}\frac{\partial C^f}{\partial z} - \frac{D}{J_w}\frac{\partial^2 C^f}{\partial z^2} = 0\ . \tag{D.218}$$

Since ϕ is constant by assumption, this equation will be invariant if

$$\frac{D}{J_w} = \text{constant} \Rightarrow \lambda := \frac{D}{V} = \text{constant} \tag{D.219}$$

as a function of J_w. This is commonly assumed to be true in the CDE.

Solution 6.4

The equation to be solved for the first layer is

$$L\frac{d\theta_1}{dt} = F - K_0\exp\big(\beta(\theta_1-\theta_0)\big) \quad , \quad \theta_1(N\Delta t) = \theta_{1i}\ , \tag{D.220}$$

where

$$F := \frac{1}{\Delta t}\int_{N\Delta t}^{(N+1)\Delta t} J_w(0,t')\,dt' \tag{D.221}$$

is the average water input flux during $N\Delta t \le t < (N+1)\Delta t$. Equation (D.220) may be rewritten as an integral

$$\frac{\Delta t}{L} = \int_{\theta_{1i}}^{\theta_{1f}} \frac{d\theta_1}{F - K_0\exp\big(\beta(\theta_1-\theta_0)\big)}\ , \tag{D.222}$$

where $\theta_{1f} = \theta_1(N+1)\Delta t$. If we let

$$y := F - K_0 \exp\big(\beta(\theta_1 - \theta_0)\big) \quad , \quad dy = \beta(y - F)d\theta_1, \tag{D.223}$$

the integral (D.222) becomes

$$\frac{\Delta t}{L} = \frac{1}{\beta F}\int_{y_i}^{y_f}\left(\frac{1}{y-F} - \frac{1}{y}\right)dy = \frac{1}{\beta F}\ln\left(\frac{y_f - F}{y_f}\frac{y_i}{y_i - F}\right) . \tag{D.224}$$

Thus, the solution for $y_f = y(\theta_f)$ is

$$y_f = \frac{F}{1-\omega} , \tag{D.225}$$

where

$$\omega := \frac{y_i - F}{y_i}\exp\left(\frac{\beta F \Delta t}{L}\right) . \tag{D.226}$$

Therefore, we may write the solution for θ_f as

$$\theta_f = \theta_0 + \frac{1}{\beta}\ln\left(\frac{F - y_f}{K_0}\right) . \tag{D.227}$$

We may repeat this for each subsequent layer, using the value of θ for each layer at $t = N\Delta t$ as the initial condition, and using the average input flux to the layer during $N\Delta t \le t < (N+1)\Delta t$ for F. In the case of layer 2, this is equal to

$$F = \overline{J_{w_2,\mathrm{in}}} = \overline{J_{w_1,\mathrm{out}}} = L(\theta_{1f} - \theta_{1i}) . \tag{D.228}$$

Solution 6.5†

The pdf $f_P(I \mid t)$ gives the probability density that the applied water will be I at time t. The two pdfs defining the climate are $f_\tau(\tau)$ and $f_h(h)$, where τ is the time interval between storms and h is the amount of water between storms. They are given by

$$f_\tau(\tau) = \omega \exp(-\omega\tau) , \tag{D.229}$$

$$f_h(h) = \frac{\lambda^\kappa h^{\kappa-1}\exp(-\lambda h)}{\Gamma(\kappa)} . \tag{D.230}$$

The probability that there will be ν storms in the time t is the same as the probability that there will be ν time intervals between 0 and t, such that the sum of the time intervals equals t. Thus,

$$f_\nu(t) = \int_0^\infty \ldots \int_0^\infty \delta\left(t - \sum_{j=1}^{\nu}\tau_j\right) f_\tau(\tau_1)\ldots f_\tau(\tau_\nu)\,d\tau_1\ldots d\tau_\nu , \tag{D.231}$$

where it has been assumed that the time intervals are independent. Equation (D.231) may be evaluated easily using the Laplace transform operation. The transform of $f_\nu(t)$ is

$$\begin{aligned} \widehat{f}_\nu(t) &= \int_0^\infty f_\nu(t)\exp(-st)\,dt \\ &= \\ &= \int_0^\infty \ldots \int_0^\infty \exp\Big(-s\sum_{j=1}^{\nu} \tau_j\Big) f_\tau(\tau_1)\ldots f_\tau(\tau_\nu)\,d\tau_1\ldots d\tau_\nu \\ &= \\ &= \prod_{j=1}^{\nu} \int_0^\infty \exp\Big(-s\tau_j\Big) f_\tau(\tau_j)\,d\tau_j \\ &= \\ &= \prod_{j=1}^{\nu} \widehat{f}_\tau(s) \Big(\frac{\omega}{\omega+s}\Big)^\nu , \end{aligned} \tag{D.232}$$

(D.233)

by (D.228) and (C.4). The inverse transform of (D.232) is evaluated using the shifting theorem (A.33) and (C.3). This produces

$$\mathcal{L}^{-1}\bigg(\Big(\frac{\omega}{\omega+s}\Big)^\nu\bigg) = \omega^\nu \exp(-\omega t)\mathcal{L}^{-1}\bigg(\Big(\frac{1}{s}\Big)^\nu\bigg) = \frac{\omega^\nu \exp(-\omega t)t^\nu}{\nu!} . \tag{D.234}$$

Similarly, the amount of water in ν storms is given by

$$f_\nu(I) = \int_0^\infty \ldots \int_0^\infty \delta\Big(I - \sum_{j=1}^{\nu} h_j\Big)\, f_h(h_1)\ldots f_h(h_\nu)\,dh_1\ldots dh_\nu . \tag{D.235}$$

The Laplace transform of (D.235) is, by (D.230),

$$\widehat{f}_\nu(s) = \int_0^\infty f_\nu(I)\exp(-sI)\,dI = \prod_{j=1}^{\nu} \widehat{f}_h(s) = \prod_{j=1}^{\nu}\Big(\frac{\lambda}{\lambda+s}\Big)^\kappa = \Big(\frac{\lambda}{\lambda+s}\Big)^{\nu\kappa} , \tag{D.236}$$

where we have used

$$\mathcal{L}\Big(\frac{\lambda^\kappa h^{\kappa-1}\exp(-\lambda h)}{\Gamma(\kappa)}\Big) = \Big(\frac{\lambda}{\lambda+s}\Big)^\kappa \tag{D.237}$$

(Abramowitz and Stegun, 1970). Finally, the inverse transform of (D.236) is

$$\begin{aligned} f_\nu(I) &= \mathcal{L}^{-1}\bigg(\Big(\frac{\lambda}{\lambda+s}\Big)^{\nu\kappa}\bigg)\lambda^{\nu\kappa} = \exp(-\lambda I)\mathcal{L}^{-1}\bigg(\Big(\frac{1}{s}\Big)^{\nu\kappa}\bigg) \\ &= \frac{\lambda^{\nu\kappa}\exp(-\lambda I)I^{\nu\kappa-1}}{\Gamma(\nu\kappa)} . \end{aligned} \tag{D.238}$$

This is the pdf defining the probability of having an amount of water I in ν storms. Thus, the conditional pdf defining the probability of having an amount of water I arriving in time t is obtained by summing over all possible storms,

$$f_P(I \mid t) = \sum_{\nu=1}^{\infty} f_\nu(I) f_\nu(t) = \sum_{\nu=1}^{\infty} \frac{\lambda^{\nu\kappa} I^{\nu\kappa-1}}{\Gamma(\nu\kappa)} \exp(-\lambda I) \frac{\omega^\nu \exp(-\omega t) t^\nu}{\nu!} . \quad \text{(D.239)}$$

Note that if we integrate over all I, we obtain

$$\begin{aligned} \int_0^\infty f_P(I \mid t)\, dI &= \sum_{\nu=1}^{\infty} \int_0^\infty \frac{\lambda^{\nu\kappa} I^{\nu\kappa-1}}{\Gamma(\nu\kappa)} \exp(-\lambda I)\, dI \frac{\omega^\nu \exp(-\omega t) t^\nu}{\nu!} \\ &= \sum_{\nu=1}^{\infty} \frac{\omega^\nu \exp(-\omega t) t^\nu}{\nu!} = 1 - \exp(-\omega t) . \end{aligned} \quad \text{(D.240)}$$

Thus, there is a probability $P(0 \mid t) = \exp(-\omega t)$ of receiving no water in time t. This derivation was outlined in Eagleson (1978).

Solution 6.6

The cumulative mass per area M crossing the plane at z is given by

$$\begin{aligned} M &= \int_0^\infty C^f(z, I)\, dI \\ &= \int_0^\infty \int_0^I C^f(0, I - I')\, f^{SS}(z, I' - \Delta W(0))\, dI' dI \\ &= \int_0^\infty \int_0^I C^f(0, I')\, f^{SS}(z, I - I' - \Delta W(0))\, dI' dI \\ &= \int_0^\infty \int_0^\infty C^f(0, I')\, f^{SS}(z, I - I' - \Delta W(0))\, dI' dI , \end{aligned} \quad \text{(D.241)}$$

where the change in the integral limits in the last line follows from the requirement that the pdf is zero for negative values of I. We now integrate over I, producing

$$\begin{aligned} M &= \int_0^\infty C^f(0, I') \int_0^\infty f^{SS}(z, I - I' - \Delta W(0))\, dI dI' \\ &= \int_0^\infty C^f(0, I') \int_{I' + \Delta W(0)}^\infty f^{SS}(z, I - I' - \Delta W(0))\, dI dI' \\ &= \int_0^\infty C^f(0, I') \int_0^\infty f^{SS}(z, X)\, dX dI' \\ &= \int_0^\infty C^f(0, I')\, dI' . \end{aligned} \quad \text{(D.242)}$$

However, the last line is just the mass input through the surface. Thus, mass is conserved in the transport volume.

Solution 6.7

The numerator of (6.20) may be written as, using (6.16) and (6.19)

$$\int_0^\infty f_P(I \mid t) f_a^f(\ell, I)\, dI = A \sum_{\nu=1}^{\infty} \frac{\Omega_\nu(t)(\eta\kappa)^{\nu\kappa}}{\Gamma(\nu\kappa)} \int_0^\infty \exp(-I(\beta/R + \eta\kappa))\, I^{\nu\kappa+\alpha-1}\, dI \ , \tag{D.243}$$

where

$$\Omega_\nu(t) := (\omega t)^\nu \frac{\exp(-\omega t)}{\nu!} \ , \quad A := \frac{1}{\alpha!}\left(\frac{\beta}{R}\right)^{1+\alpha} . \tag{D.244}$$

Using (D.39), we may write (D.243) as

$$\int_0^\infty f_P(I \mid t) f_a^f(\ell, I)\, dI = B \sum_{\nu=1}^{\infty} \Omega_\nu(t) \chi^{\nu\kappa} \frac{\Gamma(\nu\kappa + \alpha)}{\Gamma(\nu\kappa)} \ , \tag{D.245}$$

where

$$B := \frac{A}{(\beta/R + \eta\kappa)^\alpha} \ , \quad \chi := \frac{1}{1 + \beta/\eta\kappa R} . \tag{D.246}$$

Similarly, the denominator of (6.20) is equal to

$$\int_0^\infty \int_0^\infty f_P(I \mid t) f_a^f(\ell, I)\, dI dt = \frac{B}{\omega} \sum_{\nu=1}^{\infty} \chi^{\nu\kappa} \frac{\Gamma(\nu\kappa + \alpha)}{\Gamma(\nu\kappa)} . \tag{D.247}$$

The ratio of (D.245) to (D.247) is equal to (6.21).

Solution 6.8†

The travel time cdf for this special case is

$$P(t) = P_a^f(NH) \qquad \text{at } t = (N-1)\tau, \quad N = 1, 2, \ldots \ , \tag{D.248}$$

where

$$P_a^f(I) = \int_0^I f_a^f(I)\, dI \ . \tag{D.249}$$

Thus, the travel time pdf $f(t)$ is equal to the time derivative of (D.248), or

$$f(t) = \sum_{N=1}^{\infty} \Big(P_a^f(NH) - P_a^f((N-1)H)\Big)\delta\big(t - (N-1)\tau\big) \tag{D.250}$$

which is normalized.

The mean RMF is therefore

$$\mathrm{E}(\exp(-\mu t)) = \sum_{N=1}^{\infty} \Big(P_a^f(NH) - P_a^f((N-1)H)\Big) \exp\Big(-\mu\big(t - (N-1)\tau\big)\Big) \ . \tag{D.251}$$

Solution 6.9

By (6.27), the saturated hydraulic conductivity K_s is related to the reference site K_s^* by

$$K_s = \alpha^2 K_s^* \ . \tag{D.252}$$

Thus, as was shown in Problem 4.4, K_s is lognormally distributed if α is lognormally distributed, and it has the parameters

$$\mu_K = \ln(K_s^*) + 2\mu_\alpha \qquad \sigma_K = 2\sigma_\alpha \ . \tag{D.253}$$

The mode, median, and mean of K_s are therefore (Aitcheson and Brown, 1976)

$$\begin{aligned} \text{mode:} &\quad \exp(\mu_K - \sigma_K^2) = K_s^* \exp(2\mu_\alpha - 4\sigma_\alpha^2) \ , \\ \text{median:} &\quad \exp(\mu_K) = K_s^* \exp(2\mu_\alpha) \ , \\ \text{mean:} &\quad \exp(\mu_K - \sigma_K^2/2) = K_s^* \exp(2\mu_\alpha - 2\sigma_\alpha^2) \ . \end{aligned} \tag{D.254}$$

The table below summarizes the properties of the K_s distribution calculated from two α distributions. These two distributions are clearly very different, indicating that the geometric similitude scaling is not valid on the field analysed by Warrick et al., (1977).

Origin of Distribution	Properties of Computed $f(K_s)$		
	Mode [cm/hr]	Median [cm/hr]	Mean [cm/hr]
α from $\psi(\theta)$	2.72	7.71	13.0
α from $K(\theta)$	0.01	2.89	44.7

D.7 Problems in Chapter 7

Solution 7.1

The variance (7.3) may be written as

$$\frac{\mathrm{Var}_z(t)}{\sigma^2} = \sum_{\substack{i=1\\j=1}}^{N} \rho^{|i-j|} = N + 2 \sum_{\substack{i=1\\j=i+1}}^{N} \rho^{i-j} \ . \tag{D.255}$$

Making use of the well-known result from a Taylor series expansion that

$$\sum_{j=0}^{\infty} \rho^j = \frac{1}{1-\rho} \ ; \quad 0 < \rho < 1 \ , \tag{D.256}$$

we can easily make the following extensions

$$\sum_{j=N}^{\infty} \rho^j = \rho^N \sum_{j=0}^{\infty} \rho^j = \frac{\rho^N}{1-\rho} \tag{D.257}$$

$$=$$

$$\sum_{j=0}^{N} \rho^j = \sum_{j=0}^{\infty} \rho^j - \sum_{j=N+1}^{\infty} \rho^j = \frac{1-\rho^{N+1}}{1-\rho} \ . \tag{D.258}$$

Therefore,

$$\frac{\mathrm{Var}_z(t)}{\sigma^2} = N + 2 \sum_{\substack{i=1 \\ j=i+1}}^{N} \rho^{i-j} = N + 2 \sum_{i=1}^{N} \sum_{k=1}^{N-i} \rho^k$$

$$= N + 2 \sum_{i=1}^{N} \left[\sum_{k=1}^{\infty} \rho^k - \sum_{k=N-i+1}^{\infty} \rho^k \right] = N + \frac{2\rho N}{1-\rho} - \frac{2}{1-\rho} \sum_{i=1}^{N} \rho^{N-i+1} \tag{D.259}$$

$$= N + \frac{2\rho N}{1-\rho} - \frac{2}{1-\rho} \sum_{k=1}^{N} \rho^k = N + \frac{2\rho N}{1-\rho} - \frac{2(\rho - \rho^{N+1})}{(1-\rho)^2} \ .$$

Solution 7.2

At $z = \Delta z$, the expression (7.9) for the dispersivity reduces to

$$\lambda = \gamma \qquad \text{at } z = \Delta z \ . \tag{D.260}$$

As $z \to \infty$, the exponential term vanishes, and

$$\lambda \to \gamma \frac{1+\rho}{1-\rho} \qquad \text{as } z \to \infty \ . \tag{D.261}$$

Thus, the correlation coefficient determines how large the asymptotic dispersivity will be.

Solution 7.3

The second moment of the travel time (7.24) is given by

$$\mathrm{E}_z(t^2) = \mathrm{E}\left(\int_0^t \int_0^t \frac{dx}{v(x)} \frac{dy}{v(y)} \right) = \int_0^t \int_0^t \mathrm{E}\left(\frac{1}{v(x)} \frac{1}{v(y)} \right) dxdy \ . \tag{D.262}$$

However, since $1/v(z)$ is second-order stationary, then by (7.13)

$$\mathrm{E}_z(t^2) = \left(\frac{z}{V} \right)^2 + \int_0^z \int_0^{z+h} \mathrm{Cov}\left(\frac{1}{\zeta(x)}, \frac{1}{\zeta(x+h)} \right) dxdh \ . \tag{D.263}$$

Substitution of the autocorrelation function (7.16) for the covariance in (D.263) produces (7.27).

Solution 7.4

It is always possible to rotate the coordinate system so that the direction of motion is along the z-axes and $\boldsymbol{\sigma}$ is diagonal. In this case, the multivariate Gaussian (7.38) reduces to

$$f(\boldsymbol{x},t) = \prod_{j=1}^{3} \frac{1}{\sqrt{2\pi\sigma_{jj}}} \exp\left(-\frac{1}{2}\frac{(x_j - \overline{u}_j t)^2}{\sigma_{jj}}\right) , \tag{D.264}$$

where $\boldsymbol{x} = (x,y,z)$, $\overline{\boldsymbol{u}} = (0,0,\overline{u})$ and $\boldsymbol{\sigma}$ is time dependent. By direct differentiation of (D.264), we obtain

$$\frac{\partial f(\boldsymbol{x},t)}{\partial t} = \left[\sum_{j=1}^{3} \frac{1}{2}\frac{d\sigma_{jj}}{dt}\left(\frac{(x_j - \overline{u}_j t)^2}{\sigma_{jj}^2} - \frac{1}{\sigma_{jj}}\right) + \sum_{j=1}^{3} \frac{(x_j - \overline{u}_j t)\overline{u}_j}{\sigma_{jj}}\right] f(\boldsymbol{x},t) \tag{D.265}$$

$$\frac{\partial f(\boldsymbol{x},t)}{\partial x_j} = -\left[\frac{x_j - \overline{u}_j t}{\sigma_{jj}}\right] f(\boldsymbol{x},t) \tag{D.266}$$

$$\frac{\partial^2 f(\boldsymbol{x},t)}{\partial x_j^2} = \left[\frac{(x_j - \overline{u}_j t)^2}{\sigma_{jj}^2} - \frac{1}{\sigma_{jj}}\right] f(\boldsymbol{x},t) , \tag{D.267}$$

where, from (7.41),

$$D_{jj} = \frac{1}{2}\frac{d\sigma_{jj}}{dt} . \tag{D.268}$$

Substituting (D.266)–(D.268) into (D.265), we obtain

$$\frac{\partial f}{\partial t} = \sum_{j=1}^{3} D_{jj} \frac{\partial^2 f}{\partial x_j^2} - \sum_{j=1}^{3} \overline{u}_j \frac{\partial f}{\partial x_j} , \tag{D.269}$$

which is the diagonalized form of (7.39).

Solution 7.5

We may write the value of the random variable at X_j as

$$\begin{aligned} Z_j - m &= \rho(Z_{j-1} - m) + \phi s \xi_j \\ &= \rho\big(\rho(Z_{j-2} - m) + \phi s \xi_{j-1}\big) + \phi s \xi_j \\ &\;\;\vdots \\ &= \rho^{j-k}(Z_{j-k} - m) + \phi s \sum_{i=k+1}^{j} \rho^{j-i}\xi_i , \end{aligned} \tag{D.270}$$

where $\phi := \sqrt{1-\rho^2}$. Thus, by (7.15)

$$\begin{aligned}
\gamma(x_j - x_k) &= \frac{1}{2}\operatorname{Var}(Z_j - Z_k) = \frac{1}{2}\operatorname{E}\Big[\Big((Z_j - m) - (Z_k - m)\Big)^2\Big] \\
&= \frac{1}{2}\operatorname{E}\Big[\Big((1-\rho^{j-k})(Z_{j-k} - m) - \phi s \sum_{i=k+1}^{j} \rho^{j-i}\xi_j\Big)^2\Big] \\
&= \frac{s^2}{2}(1-\rho^{j-k})^2 + \frac{s^2}{2}(1-\rho^2)\sum_{i=k+1}^{j} \rho^{2(j-i)} \\
&= s^2(1-\rho^{j-k}) .
\end{aligned} \tag{D.271}$$

Similarly, the covariance and correlation functions are, by (7.13)–(7.16)

$$\begin{aligned}
\operatorname{Cov}(x_j - x_k) &= s^2\rho^{j-k} , \\
\rho(x_j - x_k) &= \rho^{j-k} = \exp\Big(-\frac{h}{L_z}\Big) ,
\end{aligned} \tag{D.272}$$

where

$$\begin{aligned}
h &= (j-k)\Delta x , \\
L_z &= -\frac{\Delta x}{\ln(\rho)} ,
\end{aligned} \tag{D.273}$$

L_z is the integral scale of Z.

Solution 7.6

No Horizontal Dispersion ($d_x = 0$)
For $d_x = 0$ the movement of the particles may be split into a constant convective part $v(x)t_k$ and a purely diffusive part $\sqrt{24d_z(x)\Delta t}\sum_{j=1}^{k}\omega_j$. We first calculate the local moments $m_1(t_k;x)$ and $m_2(t_k;x)$ for fixed x as ensemble averages denoted by $\{.\}$

$$m_1(t_k;x) = \{z(t_k;x)\} = \{v(x)t_k + \sqrt{24d_z(x)\Delta t}\sum_{j=1}^{k}\omega_j\} = v(x)t_k , \tag{D.274}$$

$$\begin{aligned}
m_2(t_k;x) &= \{z^2(t_k;x)\} \\
&= \Big\{(v(x)t_k)^2 + 2v(x)t_k\sqrt{24d_z(x)\Delta t}\sum_{j=1}^{k}\omega_j + 24d_z(x)\Delta t\Big(\sum_{j=1}^{k}\omega_j\Big)^2\Big\} \\
&= (v(x)t_k)^2 + 2d_z(x)t_k ,
\end{aligned} \tag{D.275}$$

where we used

$$\left\{\sum_{j=1}^{k} \omega_j\right\} = 0 \ , \tag{D.276}$$

$$\left\{\left(\sum_{j=1}^{k} \omega_j\right)^2\right\} = \frac{k}{12} \ . \tag{D.277}$$

We then calculate the global moments $M_1(t_k)$ and $M_2(t_k)$ with (7.60), denoting horizontal averaging by $\langle . \rangle$,

$$M_1(t_k) = \langle v(x)\rangle t_k = \langle z(t_k)\rangle \ , \tag{D.278}$$

$$M_2(t_k) = \langle v^2(x)\rangle t_k^2 + 2\langle d_z(x)\rangle t_k \tag{D.279}$$

and from this the variance of the travel depths

$$\mathrm{Var}(\boldsymbol{z}(t_k)) = M_2(t_k) - M_1^2(t_k) = 2\langle d_z(x)\rangle t_k + t_k^2(\langle v^2(x)\rangle - \langle v(x)\rangle^2) \ . \tag{D.280}$$

Noting that $\langle v^2(x)\rangle - \langle v(x)\rangle^2$ is the variance of the velocity, we may write (D.280) as

$$\mathrm{Var}(\boldsymbol{z}(t_k)) = 2\Big[\langle d_z(x)\rangle + \frac{t_k}{2}\,\mathrm{Var}(v)\Big] t_k \ . \tag{D.281}$$

Infinite Horizontal Dispersion ($d_x \to \infty$)

In contrast to the case $d_x = 0$, the increments $\Delta \boldsymbol{z}_k := \boldsymbol{z}(t_k) - \boldsymbol{z}(t_{k-1})$ are now independent and equally distributed. We may therefore first calculate the moments $M_1(\Delta t)$ and $M_2(\Delta t)$ of the increments and then apply the central limit theorem to obtain the moments of $\boldsymbol{z}(t_k) = \sum_{j=1}^{k} \Delta \boldsymbol{z}_j$ for $k \to \infty$.

For times $t < \Delta t$, the velocity of a particle is constant by the model assumptions. Therefore, the moments $M_1(\Delta t)$ and $M_2(\Delta t)$ are independent of d_x and are given by (D.278)–(D.279), and we obtain

$$\langle \Delta \boldsymbol{z}\rangle := M_1(\Delta t) = \langle v(x)\rangle \Delta t \ , \tag{D.282}$$

$$\mathrm{Var}(\Delta \boldsymbol{z}) := M_2(\Delta t) - M_1^2(\Delta t) = 2\Big[\langle d_z(x)\rangle + \frac{\Delta t}{2}\,\mathrm{Var}(v)\Big]\Delta t \tag{D.283}$$

for mean and variance of the increment $\Delta \boldsymbol{z}$. Applying the central limit theorem, we obtain mean and variance of $\boldsymbol{z}(t_k)$ for large times t_k

$$\langle \boldsymbol{z}(t_k)\rangle \xrightarrow{k\to\infty} k\langle \Delta \boldsymbol{z}\rangle = \langle v(x)\rangle t_k \ , \tag{D.284}$$

$$\mathrm{Var}(\boldsymbol{z}(t_k)) \xrightarrow{k\to\infty} k\,\mathrm{Var}(\Delta \boldsymbol{z}) = 2\Big[\langle d_z(x)\rangle + \frac{\Delta t}{2}\,\mathrm{Var}(v)\Big] t_k \ . \tag{D.285}$$

References

Abramowitz, M. and I.A. Stegun, 1970: Handbook of Mathematical Functions. Dover Publishing Co., New York.

Addiscot, T.M., 1977: A simple computer model for leaching in structured soils. *J. Soil Sci.*, **28**, 554–563

Aitcheson and Brown, 1976: The Lognormal Distribution, Cambridge University Press, Cambridge, GB.

Arfken, G., 1985: Mathematical Methods for Physicists. 3nd edition. Academic Press, New York.

Barry, D.A., J. Coves, and G. Sposito, 1988: On the Dagan model of solute transport in groundwater: Application to the Borden site. *Water Resour. Res.*, **24**, 1805–1817.

Barry, D.A. and J.C. Parker, 1987: Approximations for solute transport through porous media with flow transverse to layering. *Transp. in Por. Med.*, **2**, 65–84.

Baver, L.D., W.H. Gardner, and W.R. Gardner, 1972: Soil Physics. 4th edition, John Wiley&Sons, New York.

Biggar, J.W. and D.R. Nielsen, 1967: Miscible displacement and leaching phenomena. *Agronomy* **11**, 254–274.

Biggar, J.W. and D.R. Nielsen, 1976: Spatial variability of the leaching characteristics of a field soil. *Water Resour. Res.*, **12**, 78–84.

Black, T.A., W.R. Gardner, and G.W. Thurtell, 1969: The prediction of evaporation, drainage, and soil water storage for a bare soil. *Soil Sci. Soc. Am. Proc.*, **33**, 655–660.

Bresler, E. and G. Dagan, 1979: Solute dispersion in unsaturated heterogeneous soil at field scale. 2. Applications. *Soil Sci. Soc. Am. J.*, **43**, 467–472.

Bresler, E. and G. Dagan, 1983: Unsaturated flow in spatially variable fields. 2. Application of water flow models to various fields. *Water Resour. Res.*, **19**, 421–428.

Bresler, E. and G. Dagan, 1983: Unsaturated flow in spatially variable fields. 3. Solute transport models and their application to two fields. *Water Resour. Res.*, **19**, 429–435.

Butters, G.L., 1987: Field scale transport of bromide in an unsaturated soil. PhD Thesis, University of California, Riverside.

Butters, G.L., W.A. Jury, and F.F. Ernst, 1989: Field scale transport of bromide in an unsaturated soil. 1. Experimental methodology and results. *Water Resour. Res.*, **25**, 1575–1581.

Butters, G.L. and W.A. Jury, 1989: Field scale transport of bromide in an unsaturated soil. 2. Dispersion modeling. *Water Resour. Res.*, **25**, 1582–1588.

Carslaw, H.S. and J.C. Jaeger, 1959: Conduction of Heat in Solids. Oxford University Press, London.

Coats, K.H. and B.D. Smith, 1956: Dead end pore volume and dispersion in porous media. *Soc. Pet. Eng. J.*, **4**, 73–84.

Dagan, G., 1982: Stochastic modeling of groundwater flow by unconditional and conditional probabilities. *Water Resour. Res.*, **18**, 813–833.

Dagan, G., 1984: Solute transport in heterogenous porous formations. *J. Fluid Mech.*, **145**, 151–177.

Dagan, G., 1987: Theory of solute transport by groundwater. *Ann. Rev. Fluid Mech.*, **19**, 183–215

Dagan, G. and E. Bresler, 1979: Solute dispersion in unsaturated heterogeneous soil at field scale. 1. Theory. *Soil Sci. Soc. Am. J.*, **43**, 461:467.

Dagan, G. and E. Bresler, 1983: Unsaturated flow in spatially variable fields. 1. Derivation of models of infiltration and redistribution. *Water Resour. Res.*, **19**, 413–420.

Dagan, G., and V. Nguyen,1989: A comparison of travel time and concentration approaches to modeling transport by groundwater. *J. Contam. Hydrol.*, **4**, 79–91.

De Smedt, F. and P.J. Wierenga, 1984: Solute transfer through columns of glass beads. *Water Resour. Res.*, **20**, 225–232.

Dirac, P.A.M., 1947: The Principles of Quantum Mechanics. 3rd edition, Clarendon Press, Oxford.

Dyson, J.S. and R.E. White, 1986: The effect of irrigation rate on solute transport in soil during steady water flow. *J. Hydrol.*, **107**, 19–29.

Eagleson, P.S., 1978: Climate, soil and vegetation 1–7. *Water Resour. Res.*, **14**, 705–776.

El Abd, H., 1984: Spatial variability of the pesticide distribution coefficient. PhD Thesis, University of California, Riverside.

Ellsworth, T.E., 1989: A three dimensional field study of solute leaching through unsaturated soil. PhD Thesis, University of California, Riverside.

Ellsworth, T.E., W.A. Jury, and F.E. Ernst, 1990: A three dimensional field study of solute leaching through unsaturated soil. 1. Methodology, mass balance, and mean transport. *Water Resour. Res.*, (in review).

Elrick, D.E., J.H. Scandrett, and E.E. Miller, 1959: Tests of capillary flow scaling. *Soil Sci. Soc. Am. Proc.*, **23**, 329–332.

Freyberg, D., 1986: A natural gradient experiment on solute transport in a sand aquifer. 2. Spatial moments and advection and dispersion of nonreactive tracers. *Water Resour. Res.*, **22**, 2031–2046.

Fried, J.J., 1975: Groundwater Pollution. Elsevier Science, New York.

Gardiner, C.W., 1985: Handbook of Stochastic Methods for Physics, Chemistry and the Natural Sciences. 2nd edition, Springer-Verlag, New York.

Gel'fand, I.M. and G.E. Shilov, 1964: Generalized Functions. Vol. 1: Properties and Operations. Academic Press, New York.

Gelhar, L.W., A.L. Gutjahr, and R.L. Naff, 1979: Stochastic analysis of macrodispersion in a stratified aquifer. *Water Resour. Res.*, **15**, 1387–1397.

Gelhar, L.W. and C. Axness, 1983: Three dimensional stochastic analysis of macrodispersion in aquifers. *Water Resour. Res.*, **19**, 161–180.

Gelhar, L.W., A. Mantaglou, C. Welty, and K.R. Rehfeldt, 1985: A review of field scale physical solute transport processes in unsaturated and saturated porous media. Electric Power Research Institute, Topical Rep. EA-4190, EPRI, Palo Alto.

Gershon, N.D. and A. Nir, 1969: Effect of boundary conditions of models on tracer distribution in flow through porous mediums. *Water Resour. Res.*, **5**, 830–840.

Ghodrati, M, 1989: The influence of water application method, pesticide formulation and surface preparation method on pesticide leaching. PhD Thesis, University of California, Riverside.

Grimmett, G. and D. Welsh, 1986: Probability: An Introduction. Clarendon, Oxford.

Hamaker, J.W. and J.M. Thompson, 1972: Adsorption. In *Organic Chemicals in the Soil Environment*, Marcell Dekker Inc., New York.

Hillel, D.I., 1971: Soil and Water: Physical Principles and Processes. Academic Press, New York.

Hillel, D.I., 1980: Applications of Soil Physics. Academic Press, New York.

Hillel, D.I., 1986: Unstable flow in layered soils: A review. *Hydrolog. Proc.*, **1**, 143–147.

Himmelblau, D.M., 1970: Process Analysis by Statistical Methods. Sterling Swift Publishing Co., Manchecka, Texas.

Himmelblau, D.M. and K.B. Bischoff, 1968: Process Analysis and Simulation. John Wiley&Sons, New York.

Journel, A. and Ch.J. Huijbregts, 1978: Mining Geostatistics. Academic Press, London.

Jury, W.A., 1975: Solute travel-time estimates for tile-drained soils. 1. Theory. *Soil Sci. Soc. Am. Proc.*, **39**, 1020–1028.

Jury, W.A., 1982: Simulation of solute transport using a transfer function model. *Water Resour. Res.*, **18**, 363–368.

Jury, W.A., 1985: Spatial variability of soil physical parameters in solute migration: A critical literature review. EPRI Topical Rep. EA 4228 Electric Power Research Institute, Palo Alto.

Jury, W.A., 1988: Solute transport and dispersion. In *Flow and Transport in the Natural Environment* edited by W.L. Steffen and O.T. Denmead, 1–17, Springer Verlag, Berlin.

Jury, W.A., L.H. Stolzy, and P. Shouse, 1982: A field test of the transfer function model for predicting solute movement. *Water Resour. Res.*, **18**, 368–374.

Jury , W.A. and G. Sposito, 1985: Field calibration and validation of solute transport models for the unsaturated zone. *Soil Sci. Soc. Am. J.*, **49**, 1331–1341.

Jury, W.A., G. Sposito, and R.E. White, 1986: A transfer function model of solute movement through soil. 1. Fundamental concepts. *Water Resour. Res.*, **22**, 243–247.

Jury, W.A., H. El Abd, and M. Resketo, 1986: Field study of napropamide movement through unsaturated soil. *Water Resour. Res.*, **22**, 749–755.

Jury, W.A., D.D. Focht, and W.J. Farmer, 1987: Evaluation of pesticide ground water pollution potential from standard indices of soil-chemical adsorption and biodegradation. *J. Environ. Qual.*, **16**, 422–428.

Jury, W.A. and J. Gruber, 1989: A stochastic analysis of the influence of soil and climatic variability on the estimate of pesticide ground water pollution potential. *Water Resour. Res.*, **25**, 2465–2474.

Jury, W.A., J.S. Dyson, and G.L. Butters, 1990: A transfer function model of field scale solute transport under transient water flow. *Soil Sci. Soc. Am. J.*, **54**, 327–331.

Jury, W.A., and J. Utermann, 1991: Solute transport in layered soil: Zero and perfect correlation models. *Transp. in Por. Med.*, (in review).

Kaplan, W., 1984: Advanced Calculus. Addison-Wesley, New York.

Khan A.U.H, 1988: A laboratory test of the dispersion scale effect. PhD Thesis, University of California, Riverside.

Khan, A.U.H and W.A. Jury, 1990: A laboratory test of the dispersion scale effect in column outflow experiments. *J. Contam. Hydrol.*, **5**, 119–132.

Klute, A. and G. Wilkinson, 1958: Some tests of the similar media concept of capillary flow. *Soil Sci. Soc. Am. Proc.*, **22**, 278–281.

Kreft, A. and A. Zuber, 1978: On the physical meaning of the dispersion equation and its solution for different inital and boundary conditions. *Chem. Eng. Soc.*, **33**, 1471–1480.

Kung, S.K.-J., 1988: Preferential flow in sandy soils: Mechanisms and influences. Abstract, Amer. Soc. Agron. Nat. Meeting, Anaheim, CA.

Libardi, P.L., K. Reichardt, D.R. Nielsen, and J.W. Biggar, 1980: Some simple field methods for estimating soil hydraulic conductivity. *Soil Sci. Soc. Am. J.*, **44**, 3–6.

Lindstrom, F.T., R. Haque, V.H. Freed, and L. Boersma, 1967: Theory on the movement of some herbicides in soils: Linear diffusion and convection of chemicals in soils. *J. Env. Sci. Tech.*, **1**, 561–565.

Lindstrom, F.T., L. Boersma, and H. Gardiner, 1968: 2,4-D diffusion in saturated soils. A mathematical theory. *Soil Sci.*, **105**, 107–113.

Mantaglou, A. and L.W. Gelhar, 1987a: Stochastic modeling of large scale transient unsaturated flow systems. *Water Resour. Res.*, **23**, 37–46.

Mantaglou, A. and L.W. Gelhar, 1987b: Capillary tension head variance, mean soil moisture content, and effective specific soil moisture capacity of transient unsaturated flow in stratified soils. *Water Resour. Res.*, **23**, 47–56.

Mantaglou, A. and L.W. Gelhar, 1987c: Effective hydraulic conductivities of transient unsaturated flow in stratified soils. *Water Resour. Res.*, **23**, 57–67.

Matheron, G. and G. De Marsily, 1980: Is transport in porous media always diffusive? A counterexample. *Water Resour. Res.*, **16**, 901–907.

Miller, E.E. and R.D. Miller, 1956: Physical theory for capillary flow phenomena. *J. Appl. Phys.*, **27**, 324–332.

Naumann, E.B. and B.A. Buffham, 1983: Mixing in Continuous Flow Systems. John Wiley&Sons, New York.

Nielsen D.R. and J.W. Biggar, 1962: Miscible Displacement. 3. Theoretical considerations. *Soil Sci. Soc. Am. Proc.*, **26**, 216–221.

Nielsen, D.R., J.W. Biggar, and K.T. Erh, 1973: Spatial variability of field-measured soil-water properties. *Hilgardia*, **42**, 215–259.

Nkedi-Kizza, P., J.W. Biggar, M.Th. Van Genuchten, P.J. Wierenga, H.M. Selim, J.M. Davidson, and D.R. Nielsen, 1983: Modeling tritium and chloride-36 transport through an aggregated oxisol. *Water Resour. Res.*, **19**, 691–700.

Parker, J.C. and M.Th. Van Genuchten, 1984: Flux-averaged and volume-averaged concentrations in continuum approaches to solute transport. *Water Resour. Res.*, **20**, 866–872.

Peck, A.J., R.J. Luxmoore, and L.J. Stolzy, 1977: Effects of spatial variability in water budget modeling. *Water Resour. Res.*, **13**, 348–354.

Phythian, R., 1975: Dispersion by random velocity fields. *J. Fluid Mech.*, **67**, 145–153.

Press, W.H., B.P. Flannery, S.A. Teucholsky, and W.T. Vetterling, 1986: Numerical recipes. The art of scientific computing. Cambridge University Press, Cambridge.

Raats, P.A.C., 1973: Unstable wetting fronts in uniform and nonuniform soils. *Soil Sci. Soc. Am. J.*, **37**, 681–685.

Rao, P.S.C., D.E. Rolston, R.E. Jessup, and J.M. Davidson, 1980: Solute transport in aggregated porous media. Theoretical and experimental evaluation. *Soil Sci. Soc. Am. J.*, **44**, 1139–1146.

Rao, P.S.C., A.G. Hornsby, and R.E. Jessup, 1985: Indices for ranking the potential for pesticide contamination of groundwater. *Proc. Soil Crop Sci. Soc. Fla.*, **44**, 1–8.

Richtmyer, R.D., 1978: Principles of advanced mathematical physics. Vol. 1. Springer, New York.

Rinaldo A., A. Marani, and A. Bellin, 1989: On mass response functions. *Water Resour. Res.*, **25**, 1603–1617.

Roth, K., 1989: Stofftransport im wasserungesättigten Untergrund natürlicher, heterogener Böden unter Feldbedingungen. ETH-Diss. 8907, Zürich, Switzerland.

Roth, K., H. Flühler, and W. Attinger, 1990: Transport of conservative tracer under field conditions: Qualitative modelling with random walk in a double porous medium. In *Field-Scale Solute and Water Transport through Soil* edited by K. Roth, H. Flühler, W.A. Jury and J.C. Parker, Birkhäuser-Verlag, Basel, Switzerland.

Russo, D. and E. Bresler, 1980: Scaling soil hydraulic properties of a heterogeneous field. *Soil Sci. Soc. Am. J.*, **44**, 681–684.

Schulin, R., P.J. Wierenga, H. Flühler, and J. Leuenberger, 1987: Solute transport through a stony soil. *Soil Sci. Soc. Am. J.*, **51**, 36–42.

Schwartz, L., 1950: Théorie des Distributions. Tomes 1. Hermann Cie, Paris.

Simmons, C.S., 1982: A stochastic-convective transport representation of dispersion in one dimensional porous media systems. *Water Resour. Res.*, **18**, 1193–1214.

Simmons, C.S., 1986a: A generalization of one dimensional solute transport: A stochastic-convective flow conceptualization. Proc. of 6th Ann. AGU Front Range Branch Hydrology Days. Hydrol. Days Pub. Fort Collins, Co.

Simmons, C.S., 1986b: Scale dependent effective dispersion coefficients for one dimensional solute transport. Proc. of 6th Ann. AGU Front Range Branch Hydrology Days. Hydrol. Days Pub. Fort Collins, Co.

Small, M.J. and J.R. Mullar, 1987: Long term pollutant degradation in the unsaturated zone with stochastic rainfall infiltration. *Water Resour. Res.*, **23**, 2246–2256.

Sposito, G. and W.A. Jury, 1985: Inspectional analysis in the theory of water flow through unsaturated soil. *Soil Sci. Soc. Am. J.*, **49**, 791–798.

Sposito, G. and W.A. Jury, 1988: The solute lifetime probability density function for solute movement in the subsurface zone. *J. Hydrol.*, **102**, 503–518.

Sposito, G. and D.A. Barry, 1987: On the Dagan model of solute transport through groundwater: Foundational aspects. *Water Resour. Res.*, **23**, 1867–1875.

Taylor, G.I., 1921: Diffusion by continuous movements. *Proc. Lon. Math. Soc.*, **2**, 196–212.

Taylor, G.I., 1953: The dispersion of soluble matter flowing through a capillary tube. *Proc. Lon. Math. Soc., Ser. A*, **219**, 189–203.

Tillotson, P. and D.R. Nielsen, 1984: Scale factors in soil science. *Soil Sci. Soc. Am. J.*, **48**, 953–959.

Tompson, A.F.B., E.G. Vomvoris, and L.W. Gelhar, 1988: Numerical simulation of solute transport in randomly heterogeneous porous media: motivation, model development, and application. Tech. Report #316, R.M. Parsons Laboratory, Department of Civil Engineering, MIT, Cambridge, MA.

Utermann, J.E., J. Kladivko, and W.A. Jury, 1990: Evaluation of pesticide migration in tile-drained soils with a transfer function model. *J. Environ. Qual.*, (in press).

Valocchi, A.J., 1985: Validity of the local equilibrium assumption for modeling sorbing solute transport through homogeneous soils. *Water Resour. Res.*, **21**, 808–820.

Valocchi, A.J., 1989: Spatial moment analysis of the transport of kinetically adsorbing solutes through stratified aquifers. *Water Resour. Res.*, **25**, 273–280.

Van Genuchten, M.Th., J.M. Davidson, and P.J. Wierenga, 1974: An evaluation of kinetic and equilibrium equations for the prediction of pesticide movement through porous media. *Soil Sci. Soc. Am. Proc.*, **38**, 29–34.

Van Genuchten, M.Th. and P.J. Wierenga, 1976: Mass transfer studies in sorbing porous media. 1. Analytical solutions. *Soil Sci. Soc. Am. J.*, **40**, 473–480.

Van Genuchten, M.Th. and P.J. Wierenga, 1977: Mass transfer studies in sorbing porous media. 3. Experimental evaluation with tritium. *Soil Sci. Soc. Am. J.*, **41**, 272–278.

Van Genuchten, M.Th. and W.J. Alves, 1982: Analytical solutions of the one-dimensional convection dispersion solute transport equation. US Dept. of Agriculture. Tech. Bull. 1661.

Van Kampen, N.G., 1976: Stochastic differential equations. *Phys. Rev.*, **24**, 171–228.

Van Kampen, N.G., 1981: Stochastic processes in physics and chemistry. North-Holland Publishing Company, Amsterdam.

Warrick, A.W., G.J. Mullen, and D.R. Nielsen, 1977: Scaling soil physical properties using a similar media concept. *Water Resour. Res.*, **13**, 355–362.

Warrick, A.W. and A. Amoozegar-Fard, 1979: Infiltration and drainage calculations using spatially scaled hydraulic properties. *Water Resour. Res.*, **15**, 1116–1120.

Wierenga, P.J., 1977: Solute distribution profiles computed with steady-state and transient flow models. *Soil Sci. Soc. Am. J.*, **41**, 1050–1055.

Index

G

H

I

J

L

M

T

U

V

W

30, 19

27 14
2